QUANTUM
PSYCHOLOGY

WHAT CRITICS SAY ABOUT
ROBERT ANTON WILSON

A **SUPER-GENIUS**...He has written everything I was afraid to write
   Dr. John Lilly

One of the funniest, most incisive social critics around, and with a
positive bent, thank Goddess.
   Riane Eisler, author of *The Chalice and the Blade*

A very funny man...readers with open minds will like his books.
   Robin Robertson, *Psychological Perspectives*

Robert Anton Wilson is a dazzling barker hawking tickets to the most
thrilling tilt-a-whirls and daring loop-o-planes on the midway of higher
consciousness.
   Tom Robbins, author of *Even Cowgirls Get the Blues*

STUPID
   Andrea Antonoff

The man's either a genius or Jesus
   SOUNDS (London)

A 21st Century Renaissance Man...funny, wise and optimistic...
   DENVER POST

The world's greatest writer-philosopher.
   IRISH TIMES (Dublin)

Hilarious...multi-dimensional...a laugh a paragraph.
   LOS ANGELES TIMES

Ranting and raving...negativism...
   Neal Wilgus

**One of the most important writers working in English
today**...courageous, compassionate, optimistic and original.
   Elwyn Chamberling, author of *Gates of Fire*

Should win the Nobel Prize for INTELLIGENCE.
   QUICKSILVER MESSENGER (Brighton, England)

Wilson managed to reverse every mental polarity in me, as if I had been dragged through infinity. I was astounded and delighted.
Philip K. Dick, author of *Blade Runner*

One of the leading thinkers of the modern age.
Barbara Marx Hubbard, World Future Society

A male feminist...a simpering, pussy-whipped wimp.
Lou Rollins

SEXIST
Arlene Meyers

The **most important philosopher of this century**...scholarly, witty, hip and hopeful.
Timothy Leary

What great physicist hides behind the mask of "Robert Anton Wilson?"
NEW SCIENTIST

Does for quantum mechanics what Durrell's *Alexandria Quartet* did for Relativity, but **Wilson is funnier.**
John Gribbin, physicist

OBSCENE, blasphemous, subversive and very, very interesting.
Alan Watts

Erudite, witty and genuinely scary.
PUBLISHER'S WEEKLY

Deliberately annoying.
Jay Kinney

Misguided malicious fanaticism.
Robert Sheafer, Committee for Scientific Investigation of Claims of the Paranormal

The man's glittering intelligence won't let you rest. With each new book, I look forward to his wisdom, laced with his special brand of crazy humor.
Alan Harrington, author of *The Immortalist*

Other Books By Robert Anton Wilson

1972   Playboy's Book of Forbidden Words
1973   *Sex, Drugs and Magick: A Journey Beyond Limits
1973   The Sex Magicians
1974   *The Book of the Breast (now 'Ishtar Rising')
1975   ILLUMINATUS! (with Robert Shea)
        The Eye in the Pyramid
        The Golden Apple
        Leviathan
1977   *Cosmic Trigger I: Final Secret of the Illuminati
1978   *Neuropolitics (with T. Leary & G. Koopman)
1980   The Illuminati Papers
1980-1 The Schrodinger's Cat Trilogy
        The Universe Next Door
        The Trick Top Hat
        The Homing Pigeon
1981   Masks of the Illuminati
1983   Right Where You Are Sitting Now
1983   *The Earth Will Shake
1983   *Prometheus Rising
1985   *The Widow's Son
1986   *The New Inquisition
1987   Natural Law or Don't Put a Rubber on Your Willy
1987   *Wilhelm Reich in Hell
1988   *Coincidance: A Head Test
1988   *Nature's God
1990   *Quantum Psychology
1990   *Cosmic Trigger II: Down to Earth
1991   Chaos and Beyond
1993   *Reality Is What You Can Get Away With
1995   *Cosmic Trigger III: My Life After Death
1997   *The Walls Came Tumbling Down
1998   Everything Is Under Control
2002   *TSOG: The Thing That Ate the Constitution
2003   *TSOG: The CD
2005   *email to the universe
2006   *The Tale of the Tribe

*Published by New Falcon Publications

QUANTUM PSYCHOLOGY

*How Brain Software Programs
You and Your World*

Robert Anton Wilson

NEW FALCON PUBLICATIONS
TEMPE, ARIZONA U.S.A.

International Standard Book Number: 1-56184-071-8
Library of Congress Catalog Card Number: 92-63046

First Edition 1990
Second Printing 1993
Third Printing 1996
Fourth Printing 1999
Fifth Printing 2000
Sixth Printing 2002
Seventh Printing 2003
Eighth Printing 2004
Ninth Printing 2005

Cover Art by S. Jason Black

The paper used in this publication meets the minimum requirements of the American National Standard for Permanence of Paper for Printed Library Materials Z39.48-1984

Address all inquiries to:
NEW FALCON PUBLICATIONS
1739 East Broadway Road #1-277
Tempe, AZ 85282 U.S.A.
(or)
320 East Charleston Blvd. #204-286
Las Vegas, NV 89104 U.S.A.
website: http://www.newfalcon.com
email: info@newfalcon.com

to

Laura and John Caswell

"Rise and look around you..."

TABLE OF CONTENTS

Introductory Note

Each chapter in this book contains exercizes which will help the readers comprehend and "internalize" (learn to use) the principles of Quantum Psychology. Ideally, the book should serve as a study manual for a group which meets once a week to perform the exercizes and discuss the daily-life implications of the lessons learned.

I also employ the "scatter" technique of Sufi writers. Topics do not always appear in linear, "logical" order, but in a non-linear *psycho-logical* order calculated to produce new ways of thinking and perceiving. This technique also intends to assist the process of "internalization".

Fore-Words

An Historical Glossary

It is dangerous to understand new things too quickly.
— Josiah Warren, *True Civilization*

Some parts of this book will seem "materialistic" to many readers, and those who dislike science (and "understand" new things very quickly) might even decide the whole book has a Scientific Materialist or (they might even say) "scientistic" bias. Curiously, other parts of the book will seem "mystical" (or worse-than-mystical) to other readers and these people might decide the book has an occult — or even solipsistic — bias.

I make these gloomy predictions with great assurance, based on experience. I have heard myself called a "materialist" and a "mystic" so often that I have become wearily convinced that no matter how I change my style or "angle of approach" from one book to the next, some people will always read into my pages precisely the overstatements and oversimplifications that I have most carefully avoided uttering. This problem does not seem unique to me; something similar happens to every writer, to a greater or lesser extent. Claude Shannon proved, in 1948, that "noise" gets into every communication channel, however designed.[1]

In electronics (telephone, radio, TV etc.) *noise* takes the form of static or interference or crossed wires etc. This explains why you may hear, while looking at a football game

---

1 *The Mathematical Theory of Communication*, Claude Shannon, University of Illinois Press, 1948.

on TV, some woman interrupting a forward pass to tell her grocer how many gallons of milk she wants that week.

In print, *noise* appears primarily as "typos" — words that the printer left out, parts of sentences that land in the wrong paragraph, author's corrections that get mis-read and changed one error to a different error, etc. I have even heard of a tender love story that ended, in the author's text, "He kissed her under the silent stars," which startled some readers when it appeared in print as "He kicked her under the silent stars." (Another version of this Old Author's Tale, more amusing but less believable, claims the last line appeared as "He kicked her under the cellar stairs.")

In one of my previous books, Prof. Mario Bunge appears as Prof. Mario Munge, and I still don't know how that happened, although I suspect I deserve as much blame as the typesetter. I wrote the book in Dublin, Ireland, with an article by Prof. Bunge right in front of me, but corrected the galleys in Boulder, Colorado, in the middle of a lecture tour, without the article for reference. The quotes from Prof. Bunge appeared correctly in the book but his name appeared as Munge. I hereby apologize to the Professor (and devoutly hope that he will not appear as Munge again when this paragraph gets published — a bit of typographical noise that would insult poor old Bunge one more time and render this paragraph utterly confusing to the reader...)

In conversation, *noise* can enter due to distraction, background sounds, speech impediments, foreign accents etc. and a man saying "I just hate a pompous psychiatrist" may seem, to listeners, to have said "I just ate a pompous psychiatrist."

Semantic noise also seems to haunt every communication system. A man may sincerely say "I love fish," and two listeners may both hear him correctly, yet the two will neurosemantically file this in their brains under opposite categories. One will think the man loves to dine on fish, and the other will think he loves to keep fish (in an aquarium).

Semantic noise can even create a rather convincing simulation of insanity, as Dr. Paul Watzlavick has demonstrated in several books. Dr. Watzlavick, incidentally, got his first inkling of this psychotomimetic function of semantic noise when arriving at a mental hospital as a new staff member. He reported to the office of the Chief Psychiatrist, where he found a woman sitting at the desk in the outer office. Dr. Watzlavick made the assumption he had found the boss's secretary.

"I'm Watzlavick," he said, assuming the "secretary" would know he had an appointment.

"I didn't say you were," she replied.

A bit taken aback, Dr. Watzlavick exclaimed, "But I am."

"Then why did you deny it?" she asked.

At this point, in Dr. Watzlavick's view of the situation, the woman no longer seemed a secretary. He now classified her as a schizophrenic patient who had somehow wandered into the staff offices. Naturally, he became very careful in "dealing with" her.

His revised assumption seems logical, does it not? Only poets and schizophrenics communicate in language that defies rational analysis, and poets do not normally do so in ordinary conversation, or with the above degree of opacity. They also do it with a certain elegance, lacking in this case, and usually with some kind of rhythm and sonority.

However, from the woman's point of view, Dr. Watzlavick himself had appeared as a schizophrenic patient. You see, due to noise, she had heard a different conversation.

A strange man had approached and said, "I'm *not* Slavic." Many paranoids begin a conversation with such assertions, vitally important to them, but sounding a bit strange to the rest of us.

"I didn't say you were," she replied, trying to soothe him.

"But I am," he shot back, thereby graduating from "paranoid" to "paranoid schizophrenic" in her judgment.

"Then why did you deny it?" She asked reasonably, and became very careful in "dealing with" him.

Anybody who had experience conversing with schizophrenics will recognize how both parties in this conversation felt. Dealing with poets never has quite this much hassle.

The reader will notice, as we proceed, that this Communication Jam has more in common with many famous political, religious and scientific debates than most of us have ever guessed.

In an attempt to minimize semantic noise (knowing I cannot eliminate it entirely) I offer here a kind of historical glossary, which will not only explain some of the "technical jargon" (from a variety of fields) used in this book, but will also, I hope, illustrate that my viewpoint does not belong on either side of the traditional (pre-quantum) debates that perpetually divide the academic world.

Existentialism dates back to Søren Kierkegaard, and, in his case, represented (1) a rejection of the abstract terms beloved by most Western philosophers (2) a preference for defining words and concepts in relation to concrete individuals and their concrete *choices* in real-life situations (3) a new and tricky way of defending Christianity against the onslaughts of rationalists.

For instance, "Justice is the ideal adjustment of all humans to the Will of God" contains the kind of abstraction that existentialists regard as glorified gobbledygook. It seems to say something but if you try to judge an actual case using only this as your yardstick you will find yourself more baffled than enlightened. You need something a bit more nitty-gritty. "Justice appears, approximately, when a jury sincerely attempts to think without prejudice" might pass muster with existentialist critics, but just barely. "People use the word 'justice' to rationalize their abuse of one another" would seem more plausible to Nietzschean existentialists.

The link between Nietzsche and Kierkegaard remains a bit of historical mystery. Nietzsche followed Kierkegaard in time, but whether he ever read Kierkegaard seems uncertain; the resemblance between the two may represent pure coincidence. Nietzsche's existentialism (1) also attacked the floating abstractions of traditional philosophy and a great deal of what passes for "common sense" (e.g., he rejected the terms "good", "evil", "the real world", and even "the ego") (2) also preferred concrete analysis of real-life situations, but emphasized *will* where Kierkegaard had emphasized *choice*, and (3) attacked Christianity, rather than defending it.

Briefly — too briefly, and therefore somewhat inaccurately — when we decide on a course of action and convince ourselves or others that we have "reasoned it all out logically," existentialists grow suspicious. Kierkegaard would insist that you made the *choice* on the basis of some "blind faith" or other (faith in Christianity, faith in Popular Science articles, faith in Marx...etc.) and Nietzsche would say that you as a biological organism *will* a certain result and have "rationalized" your biological drives. Long before Godel's Proof in mathematics, existentialism recognized that we never "prove" any proposition *completely* but always stop somewhere short of the infinite steps required for a total logical "proof" of anything; e.g., the abyss of infinity opens in attempting to prove "I have x dollars in the bank" as soon as one questions the concept of "having" something. (I think I "have" a working computer but I may find I "have" a non-working computer at any moment.)

"George Washington served two terms as President" seems "proven" to the average person when a Standard Reference Book "confirms" it; but this "proof" requires faith in Standard References — a faith lacking in many "revisionist" theories of history.

Sartre also rejected abstract logic, and emphasized *choice*, but had a leaning toward Marxism and went further than Kierkegaard or Nietzsche in criticizing terms without concrete referents. For instance, in a famous (and typical) passage, Sartre rejects the Freudian concept of "latent homosexuality" on the grounds that we may call a man homosexual if he performs homosexual *acts* but that we abuse the language when we assume an unobservable "essence of homosexuality" in those who do *not* perform homosexual *acts*.

Because of his emphasis on choice, Sartre also denied that we can call a man a homosexual (or a thief, or saint, or an antisemite etc.) except at a date. "Mary had a lesbian affair last year," "John stole a candy bar on Tuesday," "Robin gave a coin to a beggar on three occasions," "Evelyn said something against Jewish landlords two years ago" seem legitimate sentences according to Sartre, but implying an "essence" to these people appears fictitious. Only after a man or woman has died, he claimed, can we say definitely, "She was homosexual," "He was a thief," "He was charitable," "She was an antisemite", etc. While life and choice remain, Sartre holds, all humans lack "essence" and can change suddenly. (Nietzsche, like Buddha, went further and claimed that we lack "ego" — i.e., one unchanging "essential" self.)

One summary of existentialist theory says "Existence precedes essence." That means that we do not have an inborn metaphysical "essence", or "ego", such as assumed in most philosophy.[1] *We exist first and we perforce make choices, and, trying to understand or describe our existential choices, people attribute "essences" to us, but these "essences" remain labels — mere words.*

Nobody knows how to classify Max Stirner — a complex thinker who has strange affinities with atheism, anarchism, egotism, Zen Buddhism, amoralism, existentialism, and even Ayn Rand's Objectivism. Stirner also disliked non-referential abstractions (or "essences") and called them "spooks", a term for which I have a perhaps inordinate fondness.[2] My use of this term does not indicate a whole-hearted acceptance of Stirner's philosophy (or anti-philosophy), any more than my use of existential terms indicates total agreement with Kierkegaard, Nietzsche or Sartre.

Edmund Husserl stands midway between *existentialism* and *phenomenology*. Rejecting traditional philosophy as utterly as the existentialists, Husserl went further and

---

[1] Nor does an iron bar possess the "essence" of "hardness." It merely seems hard to humans, but might seem comparatively soft or pliable to a muscular 500-pound gorilla.

[2] "Spooks" does not appear in Stirner's German, of course. We actually owe this delightful term to Stirner's translator, Stephen Byington.

rejected all concepts of "reality" except the experiential (phenomenological). If I see a pink elephant, Husserl would say, the pink elephant belongs to the field of human experience as much as the careful measurements made by a scientist in a laboratory (although it occupies a different area of human experience and probably has less importance for humanity-in-general, unless I write a great poem about it....)

Husserl also emphasized the *creativity* in every act of perception (i.e., the brain's role as instant interpreter of data, something also noted by Nietzsche) and thus has had a strong influence on sociology and some branches of psychology.

Jan Huizinga, a Dutch sociologist, studied the game element in human behavior, and noted that we live by *game rules* which often have never risen to the level of conscious speech. In other words, we not only interpret data as we receive it; we also, quickly and unconsciously, "fit" the data to pre-existing axioms, or game-rules, of our culture (or our sub-culture).

For instance:

A cop clubs a man on the street. Observer A sees Law and Order performing their necessary function of restraining the violent with counter-violence. Observer B sees that the cop has white skin and the man hit has black skin, and draws somewhat different conclusions. Observer C arrived earlier and noted that the man pointed a gun at the cop before being clubbed. Observer D hears the cop saying "Stay away from my wife" and has a fourth view of the "meaning" of the situation. Etc.

Phenomenological sociology owes a great deal to Husserl and Huizinga, and to Existentialism. Denying abstract or Platonic "reality" (singular) the social scientists of this school recognize only social realities (plural) defined by human interactions and game-rules, and limited by the computational abilities of the human nervous system.

Ethnomethodology, largely the creation of Dr. Charles Garfinkle, combines the most radical theories of modern anthropology and phenomenological sociology. Recognizing social realities (plural), which it calls *emic realities*, ethnomethodology shows how every human perception, *including the perceptions of social scientists who think they can study society "objectively"*, always contains the limits, the defects and the unconscious prejudices of the *emic reality* (or social game) of the observer.

Phenomenologists and ethnomethodologists sometimes acknowledge an *etic reality* which is like unto the old-fashioned "objective reality" of traditional (pre-existentialist) philosophy and the ancient superstitions which have by now become "common sense". However, they point out that we cannot say anything meaningful about *etic* reality, because anything we can say has the structure of our *emic* reality — our social game rules (especially our language game) — built into it.

If you wish to deny this, please send me a complete description of etic reality, without using words, mathematics, music or other forms of human symbolism. (Send it express. I have wanted to see it for decades.)

Existentialism and phenomenology have not only influenced some social scientists but many artists and quite a few social activists or radicals. Both, however, have had bad repute among academic philosophers and their influence on the physical sciences has not received much acknowledgment. We shall now trace that influence.

Pragmatism has a family resemblance to existentialism and phenomenology and arose out of the same social manifold. This philosophy, or method, derives chiefly from William James — a man so complex that his books land in the philosophy section of some bookstores and libraries, the psychology section elsewhere, and sometimes even appear in the religion section. Like existentialism, pragmatism rejects spooky abstractions and most of the vocabulary of traditional philosophy.

According to pragmatism, ideas have meaning only in concrete human situations, "truth" as abstraction has no meaning at all, and the best we can say of any theory consists of, "Well, this idea seems to work, at least for the time being."

Instrumentalism *a la* John Dewey follows pragmatism in general, but especially emphasizes that the *validity* or *utility* of an idea — we have gotten rid of "truth", remember? — derives from the instruments used in testing the idea, and will change as instruments improve.

Like the other theories discussed thus far, Instrumentalism has had more direct influence on social science (and educational theory) than on physical science, although vastly influenced *by* physical science.

Operationalism, created by Nobel physicist Percy W. Bridgman, attempts to deal with the "common sense" objections to Relativity and Quantum Mechanics, and owes a great deal to pragmatism and instrumentalism. Bridgman explicitly pointed out that "common sense" derives unknowingly from some tenets of ancient philosophy and speculation — particularly Platonic Idealism and Aristotelian "essentialism" — and that this philosophy assumes many axioms that now appear either untrue or unprovable.

Common sense, for instance, assumes that the statement "The job was finished in five hours" can contain both absolute truth and objectivity. Operationalism, however, following Einstein (and pragmatism) insists that the only meaningful statement about that measurement would read *"While I shared the same inertial system* as the workers, my watch indicated an interval of five hours from start to finish of the job."

The contradictory statement, "The job took six hours" then seems, not false, but equally true, if the observer took the measurement from another inertial system. In that case, it should read, *"While observing the workers' inertial system from my spaceship (another inertial system moving away from them),* I observed that my watch showed an interval of six hours from start to finish of the job."

Operationalism has had a major influence on the physical sciences, a lesser influence on some social sciences, and appears largely unknown to, or rejected by, academic philosophers, artists, humanists etc. Oddly, many of these people, who dislike operationalism as a "cold, scientific" approach, have no similar objection to existentialism or phenomenology.

This seems strange to me. I regard existentialism and phenomenology as the application to human relations of the same critical methods that operationalism applies to the physical sciences.

The Copenhagen Interpretation of quantum physics, created by Niels Bohr (another Nobel winner), says much the same as operationalism, in even more radical language. According to Bohr, "common sense" and traditional philosophy both have failed to account for the data of quantum mechanics (and of Relativity) and we need to speak a new language to understand what physics has discovered.

The new language suggested by Bohr eliminates the same sort of abstractions attacked by existentialism and tells us to define things in terms of human operations, just like pragmatism and operationalism. Bohr admitted that both the existentialist Kierkegaard and the pragmatist James had influenced his thinking on these matters. (Most scientists oddly remain ignorant of this "philosophic" background of operationalism and just regard the operational approach as "common sense" — just as non-scientists regard Platonic and Aristotelian metaphysics as "common sense".)

General Semantics, the product of Polish-American engineer Alfred Korzybski, attempted to formulate a new non-Aristotelian logic to remove the "essentalist" or Aristotelian game-rules from our neurolinguistic reactions (speech and thinking) and re-align our brain software with the existentialist and phenomenological concepts of the above systems and especially of quantum mechanics. E-Prime (English without the word "is"), created by D. David Bourland, Jr., attempts to make the principles of General Semantics more efficient and easier to apply. I owe great debts to both Korzybski and Bourland.

General Semantics has influenced recent psychology and social science greatly but has had little effect on physical sciences or education and virtually no effect on the problems it attempted to alleviate — i.e., the omnipresence of unacknowledged bigotry and unconscious prejudice in most human evaluations.

Transactional psychology, based largely on the pioneering research concerning human perception conducted at Princeton University in the 1940s by Albert Ames, agrees with all the above systems that we cannot know any abstract "Truth" but only relative truths (small t, plural) derived from our *gambles* as our brain makes models of the ocean of new signals it receives every second.

Transactionalism also holds that we do not passively receive data from the universe but actively "create" the form in which we interpret the data as fast as we receive it. In short, *we do not re-act to information but experience transactions with information.*

Albert Camus in *The Rebel* refers to Karl Marx as a religious prophet "who, due to a historical misunderstanding, lies in the unbelievers' section of an English cemetery."

I assert that, due to another historical misunderstanding, operationalism and Copenhagenism have remained mostly the "property" of physicists and others in the "hard sciences", while existentialism and phenomenology have gained acceptance mostly among literary humanists and only slightly among social scientists. The viewpoint of this book combines elements from both traditions, which I think have more that unifies them than separates them.

I also assert a great unity between these traditions and radical Buddhism, but I will allow that to emerge gradually in the course of my argument.

For now, I have said enough to counteract most of the noise that might otherwise distort the messages I hope to convey. This book does not endorse the Abstract Dogmas of either Materialism or Mysticism; it tries to confine itself to the nitty-gritty real-life contexts explored by existentialism, operationalism and the sciences that employ existentialist-operationalist methods.

PART ONE

How Do We Know What We Know, If We Know Anything?

I do not pretend to tell what is absolutely true,
but what I think is true.
— Robert Ingersoll, *The Liberty of Man, Woman and Child*

You can see the above illustration two different ways. Can you see it both ways at the same time, or can you only change your mental focus rapidly and see it first one way and then the other way, in alteration?

"I detect a Ubangi in the fuel supply."   — W. C. Fields

ONE

A Parable About a Parable

A young American named Simon Moon, studying Zen in the *Zendo* (Zen school) at the New Old Lompoc House in Lompoc, California, made the mistake of reading Franz Kafka's *The Trial*. This sinister novel, combined with Zen training, proved too much for poor Simon. He became obsessed, intellectually and emotionally, with the strange parable about the door of the Law which Kafka inserts near the end of his story. Simon found Kafka's fable so disturbing, indeed, that it ruined his meditations, scattered his wits, and distracted him from his study of the *Sutras*.

Somewhat condensed, Kafka's parable goes as follows:

> A man comes to the door of the Law, seeking admittance. The guard refuses to allow him to pass the door, but says that if he waits long enough, maybe, someday in the uncertain future, he might gain admittance. The man waits and waits and grows older; he tries to bribe the guard, who takes his money but still refuses to let him through the door; the man sells all his possessions to get money to offer more bribes, which the guard accepts — but still does not allow him to enter. The guard always explains, on taking each new bribe, "I only do this so that you will not abandon hope entirely."
>
> Eventually, the man becomes old and ill, and knows that he will soon die. In his last few moments he summons the energy to ask a question that has puzzled him over the years. "I have been told," he says to the guard, "that the Law exists for all. Why then does it happen that, in all the years I have sat here waiting, nobody else has ever come to the door of the Law?"

"This door," the guard says, "has been made only for you. And now I am going to close it forever." And he slams the door as the man dies.

The more Simon brooded on this allegory, or joke, or puzzle, the more he felt that he could never understand Zen until he first understood this strange tale. If the door existed only for that man, why could he not enter? If the builders posted a guard to keep the man out, why did they also leave the door temptingly open? Why did the guard close the previously open door, when the man had become too old to attempt to rush past him and enter? Did the Buddhist doctrine of *dharma* (law) have anything in common with this parable?

Did the door of the Law represent the Byzantine bureaucracy that exists in virtually every modern government, making the whole story a political satire, such as a minor bureaucrat like Kafka might have devised in his subversive off-duty hours? Or did the Law represent God, as some commentators claim, and, in that case, did Kafka intend to parody religion or to defend its divine Mystery obliquely? Did the guard who took bribes but gave nothing but empty hope in return represent the clergy, or the human intellect in general, always feasting on shadows in the absence of real Final Answers?

Eventually, near breakdown from sheer mental fatigue, Simon went to his *roshi* (Zen teacher) and told Kafka's story of the man who waited at the door of the Law — the door that existed only for him but would not admit him, and was closed when death would no longer allow him to enter. "Please," Simon begged, "explain this Dark Parable to me."

"I will explain it," the roshi said, "if you will follow me into the meditation hall."

Simon followed the teacher to the door of the meditation hall. When they got there, the teacher stepped inside quickly, turned, and slammed the door in Simon's face.

At that moment, Simon experienced Awakening.

Exercizes

1. Let every member of the group try to explain or interpret Kafka's parable and the Zen Master's response.

2. Observe whether a consensus emerges from this discussion or each person finds a personal and unique meaning.

TWO

The Problem of "Deep Reality"

According to Dr. Nick Herbert's excellent book, *Quantum Reality,* the majority of physicists accept Niels Bohr's "Cophenhagen Interpretation" of quantum mechanics. (We will later examine the ideas of physicists who reject Copenhagenism and have other views.) According to Dr. Herbert, the Cophenhagen view means that "there is no deep reality."

Since we will soon find reasons to avoid the "is" of identity, and other forms of "is", let us reformulate that in more operational language — language that does not assume we can know what things metaphysically "are" or "are not" (their invisible "essences") but only that we can describe what we phenomenologically experience. The Copenhagen Interpretation then means, not that there "is" no "deep reality", but that scientific method can never experimentally locate or demonstrate a "deep reality" that explains all other relative (instrumental) "realities".

Dr. David Bohm, however, states it this way: "The Copenhagen view denies that we can make statements about actuality." This says something more than Dr. Herbert's formulation, if you chew on it a bit.

Both Dr. Herbert and Dr. Bohm reject the Copenhagen view. Dr. Herbert has even called Copenhagenism "the Christian Science school of physics." Like Dr. Bohm, Dr. Herbert — a good friend of mine — believes that physics *can* make statements about actuality.

I agree. But I limit "actuality" to that which humans or their instruments can detect, decode and transmit. "Deep reality" lies in another area entirely — the area of philosophy and/or "speculation." Thus, Dr. Richard Feynman said to Dr. Bohm of his recent book, *Wholeness and the Implicate Order,* "Brilliant philosophy book — but when are you going to write some physics again?"

I will defend Dr. Bohm (and Dr. Herbert) later. For the present, *actuality* in this book means something that humans can experience and "deep reality" means something that we can only make noises about. Science, like existentialism, deals with what humans can experience, and "deep reality" belongs to the pre-existential Platonic or Aristotelian philosophers.

We can only make noises about "deep reality" — we cannot make meaningful (testable) statements about it — because that which lies outside existential experience lies outside the competence of human judgment. No scientific board, no judge, no jury and no Church can *prove* anything about "deep reality", or even *disprove* anything about it. We cannot demonstrate that it has temperature or does not have temperature, that it has mass or does not have mass, that it includes one God or many Gods or no God, that it smells red or that it sounds purple, etc. We can make noises, to say that again, but we cannot produce non-verbal or phenomenological data to give meaning to our noises.

This rejection of speech about "deep reality" parallels the Heisenberg Uncertainty Principle, which in one form states that we can never measure the momentum and velocity of the same particle at the same time. It also parallels Einstein's Relativity, which says we can never know the "true" length of a rod but only the various lengths — plural — measured by various instruments in various inertial systems by observers who may share the same inertial system with the rod or may measure it from the perspective of another inertial system. (Just as we can never know the "true" time interval between two events, but only the different times — plural — measured from different inertial systems.) It also parallels the Ames demonstrations in perception psychology, which showed that *we do not perceive "reality"* but receive signals from the environment which we organize into guesses so fast that we do not even observe ourselves guessing.

Such "axioms of impotence," as somebody once called them, do not predict the future in the ordinary sense — we know that the future can always surprise us. Limitations of this sort in science merely mean that scientific method cannot, by definition, answer certain questions. If you want answers to those kinds of questions, you must go to a theologian or occultist, and the answers you will get there will not satisfy those who believe in other theologians or occultists, or those who don't believe in such Oracles at all, at all.

An elementary example: I can give a physicist, or a chemist, a book of poems. After study, the scientist can report back that the book weighs x kilograms, measures y centimeters in thickness, has been printed with ink having a certain chemical formula and bound with glue having another chemical formula etc. But scientific study cannot answer the question, "Are these good poems?" (Science in fact cannot answer any questions with "is" or "are" in them, but not all scientists realize that yet.)

So, then, the statement that *we cannot find* (or demonstrate to others) one "deep reality" (singular) that explains all the relative realities (plural) measured by our instruments — *and by our nervous system, the instrument that "reads" (interprets) all other instruments* — does not mean the same as the statement "there is no deep reality." Our inability to find one deep reality registers a demonstrable fact about scientific method and human neurology, while the statement "there *'is'* no deep reality" offers a metaphysical opinion about something we cannot test scientifically or experience existentially.

In short, we can know what our instruments and brains tell us (but we cannot know if our instruments and brains have reported accurately until other researchers duplicate our work...)

What our instruments and brains tell us consists of relative "realities" or cross-sections of "realities". A thermometer, for instance, does not measure length. A yardstick does not measure temperature. A voltmeter tells

us nothing about gas pressure. Etc. A po€
register the same spectrum as a banker. An F
not perceive the same world as a New York cat

The notion that we can find "one deep reality"
underlying all these relative instrumental/neurological
"realities" rests upon certain axioms about the universe,
and about the human mind, which seemed obvious to our
ancestors, but now seem either flatly untrue or — even
worse — "meaningless".

I had better explain "meaninglessness". To the scientists,
especially of the Cophenhagen persuasion, an idea seems
meaningless if we cannot, even in theory, imagine a way of
testing it. For instance, most scientists could classify as
"meaningless" the following three propositions:

1. The frammis goskit distims the blue doshes on round
   Thursdays.
2. All living beings contain souls which cannot be seen
   or measured.
3. God told me to tell you not to eat meat.

Try to imagine how one would prove, or disprove, these
statements on the level of experience or experiment. First,
you have to find goskits, blue doshes, souls and "God"
and then get them into the laboratory; then you have to
figure out how to measure them, or detect signals from
them, or somehow demonstrate that you at least have the
right goskits or the right "God", etc.

Stop and think about that. You will now, hopefully, see
why such propositions appear "meaningless" compared to
a statement like "Water boils at 45 degrees Fahrenheit at
sea level on this planet," which easily lends itself to testing
(and refutation) or "I feel like shit," which probably
contains truth to the speaker but always remains
problematical (but not "meaningless") to the listeners, who
know the speaker has described a common human feeling,
but do not know whether he means what he says or has
some motive for deceiving them. "I feel like shit" may
function as what Dr. Eric Berne called a Wooden Leg
Game — the attempt to shirk responsibility by feigning
incapacity.

Let us consider other untestable ideas where we can at least imagine a test, but at present lack the technology to perform the test. ("I feel like shit" may fall into this category.) Some refer to this equally enigmatic class of propositions as "indeterminate" rather than purely "meaningless". The following statements appear indeterminate:

1. Barnard's star has one or more planets circling it.
2. Homer was actually two poets writing in collaboration.
3. The first settlers of Ireland came from Africa.

We cannot "see" Barnard's star clearly enough to prove or disprove the first assertion, but probably will "see" it clearly enough to make a decision after the space telescope goes into orbit. (From Earth we can see frequent occlusions of Barnard's star which have led many astronomers to suspect our view periodically gets blocked by orbiting planets, but this deduction remains a guess as of the date I write this.) People can argue about Homer forever, but nobody will prove their case until some breakthrough in technology occurs (e.g., computer analysis of word choices may determine if a manuscript had one author or two, or we might invent a time machine...) Some day archeology may advance to the point of identifying the first inhabitants of Ireland, but now we can only infer that perhaps some came from Africa.

Thus, where Aristotelian logic assumes only the two classes "true" and "false", post-Copenhagenist science tends to assume four classes, although only Dr. Anatole Rapoport has stated the matter this clearly — "true", "false", "indeterminate" (not yet testable), and "meaning-less" (forever untestable). Some logical positivists also refer to "meaningless" statements as "abuse of language"; Nietzsche simply called them "swindles". Korzybski described them as "noises", a term I've already borrowed.

Among the propositions about the universe which underlie the "one deep reality" fallacy, one can mention the concept of the universe as a static thing, where current research seems to indicate that conceiving it as an active process fits the data better. A static thing or block-like entity can have one "deep reality" but a process has changing trajectories, evolution, Bergsonian "flux" etc. E.g., *if primates had one "deep reality" or Aristotelian "essence" we could not distinguish Shakespeare from a chimpanzee.*

(Our inability to distinguish certain Fundamentalist preachers from chimpanzees does not contradict this.)

"One deep reality" also implies the idea of the universe as a simple two-decker affair made up of "appearances" and *one* "underlying reality", like a mask with a face behind it. Modern research, however, indicates an indefinite series of appearances on different levels of instrumental magnification and finds no *one* "substance" or "thing" or "deep reality" that underlies all the different appearances reported by different classes of instruments. E.g., traditional philosophy and common sense assume that the hero and the villain have different "essences", as in melodrama (the villain may wear the mask of virtue, but we know he "is really" a villain); but modern science pictures things in flux, and flux in things, so solid becomes gas and gas becomes solid again, just as hero and villain become blurred and ambiguous in modern literature or Shakespeare.

One model, or reality-tunnel, never "wears a crown," so to speak, and sits in royal splendor above all the others. Each model has its own uses in its own appropriate area. "A good poem" has no meaning in science, but has many, many meanings for poetry-lovers — a different meaning, in fact, for each reader...

In short, "one deep reality" seems, to this view, as absurd as "one correct instrument," or the medieval "one true religion"; and preferring, say, the wave model of "matter" to the particle model seems as silly as claiming the thermometer tells more of the truth than the barometer.

Pauline Kael always hates the movies I love, but this does not mean one of us has a defective "good film detector." It merely means that we live in different emic realities.

Perhaps we have gone a bit further than the strict opera-
tionalist would like. We have not only implied that the
"physical truth" does not possess more indwelling
"deepness" than the "chemical truth", or the "biological
truth", or even the "psychiatric truth", and that all these
emic realities have uses *in their own fields,* but we open the
possibility that "existential truth" or "phenomenological
truth" (the truths of experience) have as much "depth"
(and/or "shallowness") as any organized scientific (or
philosophic) truths.

Thus, radical psychologists ask us: does not the "reality"
of schizophrenia or art remain "real" to those in
schizophrenic or artistic states, however senseless these
states appear to the non-schizophrenic or non-artistic?
Anthropologists even ask: do not the emic realities of other
cultures remain existentially real to those living in those
cultures, however bizarre they may seem to the Geriatric
White Male hierarchy that defines official "reality" in our
culture?

At the end of the 18th Century, science believed the sun
"is" a burning rock. (Now we model it as a nuclear
furnace.) William Blake, the poet, denied that the sun
"really was" a rock and claimed it "was" a band of angels
singing, "Glory, Glory, Glory to the Lord God Almighty."
Phenomenology will only say that the scientific gloss
appears useful to science, at a date, and the poetic gloss
appears useful to poets, or to some poets. This point seems
perfectly clear if one conspicuously avoids the "is of identity," as I just did, but opens a debate that spirals down-
ward to Chaos and Nonsense if one rewrites it as "the sun
is a rock, or a furnace, to science, but is also a band of
angels, to certain kinds of poets." Try debating that formulation for a while, and you'll understand why physicists
began to seem a bit mad when arguing "matter is waves
but it is also particles" (before Bohr taught them to say,
"We can model matter as waves or model it as particles, in
different contexts.")

It seems, then, from both operational and existential viewpoints that "isness" statements have no meaning, especially if they fall into such types as:

1. Physics is real; poetry is nonsense.
2. Psychology is not a true science.
3. There is only one reality, and my church (culture/ field of science/political Ideology etc.) knows all about it.
4. People who disagree with this book really are a bunch of jerks.

Nonetheless, it seems that, because the meaninglessness of all "isness" statements has not been generally recognized, many physicists confuse themselves and their readers by saying "There is no deep reality" (or even worse, "There is no such thing as reality." I have actually seen the latter in print, by a distinguished physicist, but out of mercy I won't mention his name.)

Quite similar to this confusion in quantum mechanics, popularizers of Transactional psychology — and, even more, popularizers of the Oriental philosophies that resemble Transactional psychology — often tell us that "Reality doesn't exist" or "You create your own reality." These propositions cannot be proven, and cannot be refuted either — a more serious objection to them than their lack of proof, since science now recognizes that irrefutable propositions have no operational or phenomenological "meaning".

Thus, "Whatever happens, however tragic and horrible it seems to us, serves the greater good, or God wouldn't let it happen" — a very popular idea, especially among those who have endured terrible grief — may serve a therapeutic function for those in great emotional pain, but also, alas, it contains the classic trait of purely meaningless speech. No possible evidence could refute it, since evidence falls into the category of "how things seem to us," and the statement refuses to address that category.

"You create your own reality" has the same irrefutable and untestable character, and hence also fits the class of meaningless speech, or Stirner's "spooks" (or Nietzsche's "swindles" or Korzybski's "noises").

What the popularizers should say, if they aimed at accuracy, would take a more limited and existential form. You create your own *model* of reality, or you create your own *reality-tunnel* (to borrow a phrase from the brilliant, if much maligned, Dr. Timothy Leary), or (as they say in sociology) you create your own *gloss* of the "realities" you encounter. Each of these formulations refer to definite and specific experiences in space-time, which easily confirm themselves in both daily-life demonstration and in controlled laboratory experiments on perception.

Our young/old woman in the drawing at the beginning of Chapter One represents one easy daily-life illustration. It requires a huge leap of metaphysics to proceed from this, or from laboratory demonstrations of the creativity in every act of perception, or from the paradoxes of quantum mechanics, to the resonant (but meaningless) proclamation that "you create your own reality."

So: the first point of resemblance between quantum mechanics and brain software — the first step in creating what I have dared to call Quantum Psychology — lies in recognizing the fact that the study of both "matter" and "mind" leads us to question normal notions of "reality".

The second point of resemblance lies in the fact that such questioning can easily degenerate into sheer gibberish if we do not watch our words very carefully. (And, I have learned, even if we do watch our words very carefully, some people will read carelessly and still take away a message full of the gibberish we have tried to avoid.)

Consider the following two propositions:

1. My boss is a male chauvinist drunk, and this is making me sick.
2. My secretary is an incompetent, whining bitch, and I have no choice but to fire her.

Both of these represent mental processes occurring thousands of times a day in modern business.

Both of them also appear as "abuse of language" or mere "noise" according to the modern scientific attitude presented in this book. If we imagine these sentences spoken aloud by persons in therapy, different types of psychologists would "handle" them in different fashions, but Rational-Emotive Therapists, following Dr. Albert Ellis, would force the patient to restate them in accord with the same principles urged in this chapter.

In that case, the statements would emerge, translated out of the Aristotelian into the existential, as:

1. I perceive my boss as a male chauvinist drunk, and right now I do not (or will not) perceive or remember anything else about him, and framing my experience this way, ignoring other factors, makes me feel unwell.
2. I perceive my secretary as an incompetent, whining bitch, and right now I do not (or will not) perceive or remember anything else about her, and framing my experience this way, ignoring other factors, inclines me to make the choice of firing her.

This reformulation may not solve all problems between bosses and secretaries, but it moves the problems out of the arena of medieval metaphysics into the territory where people can *meaningfully take responsibility for the choices they make.*

Exercizes

1. Let each member of the group classify each of the following propositions as meaningful or meaningless.
    A. I hauled the garbage out this morning.
    B. God appeared to me this morning.
    C. I saw a UFO this morning.
    D. This table top measures two feet by four feet.
    E. Space becomes curved in the vicinity of heavy masses, such as stars.
    F. Space does not become curved at all; light simply bends in the vicinity of heavy masses, such as stars.
    G. Defendants are innocent until the jury pronounces them guilty.
    H. The umpire's decision is binding.

I. "History is the march of God through the world."
   (Hegel)
J. In the act of conception, the male and female each
   contribute 23 chromosomes.
K. The devil made me do it.
L. My unconscious made me do it.
M. Conditioned reflexes made me do it.
N. A church is the house of God.
O. Anybody who criticizes the government is a traitor.
P. Abraham Lincoln served as President between 1960
   and 1968.

2. Where disagreements arise, attempt to avoid conflict
(quarrel) and seek to understand why disagreements must
arise in judging some of these propositions.

THREE

Husband/Wife & Wave/Particle Dualities

By the way, I have no academic qualifications to write about Quantum Mechanics at all, but this has not prevented me from discussing the subject quite cheerfully in four previous books.

Some readers may wonder where I get my *chutzpah*. After all, most physicists claim that the principles of Quantum Mechanics contain problems (or paradoxes) so abstruse and recondite that it requires a college degree in advanced mathematics to understand the subject at all. I first began to doubt that notion after a novel of mine, *Schroedinger's Cat* — the first of my books to deal entirely with quantum logic — received a very favorable review in *New Scientist*, by a physicist (John Gribbin) who claimed that I must also have a degree in advanced physics to have written the book. In fact, I do not have any degree in physics. (All I had of physics at university consisted of Newtonian mechanics, optics, light, electromagnetism and a mere survey course on the ideas of Relativity and Quantum Theory.)

If I seem to understand quantum logic fairly well, as other physicists besides Dr. Gribbin have asserted, this results from the fact that Transactional psychology, the study of how the brain processes data — a field in which I do hold some academic qualification — contains exactly the same weirdness that has made the quantum universe infamous. In fact, I might even say that *the study of brain science will prepare one for quantum theory better than the study of classical physics would.*

This may surprise many, including the physicists who claim that quantum uncertainty only applies to the subatomic world and that in ordinary affairs "we still live in a Newtonian universe." This book dares to disagree with that accepted wisdom; I take exactly the opposite position. My endeavor here will attempt to show that the celebrated "problems" and "paradoxes" and the general philosophical enigmas of the quantum world appear also in daily life.

For instance, the illustration at the beginning of Chapter One — which you can see as a young woman or as an old lady — demonstrates a fundamental discovery of perception psychology. This discovery appears in many different formulations, in various books, but the simplest and most general statement of it, I think, goes like this: *perception does not consist of passive reception of signals but of an active interpretation of signals.* (Or: perception doesn't consist of passive re-actions but of active, creative trans-actions.)

The same law appears, in quantum theory, in different words, but most commonly physicists state it as "the observer cannot be left out of the description of the observation." (Dr. John A. Wheeler goes further and says the observer "creates" the universe of observation.) I will endeavor to show that the similarity of these principles derives from a deeper similarity that unites quantum mechanics and neuroscience with each other (and with certain aspects of Oriental philosophy).

Similarly, close relatives of such quantum monsters as Einstein's Mouse, Schroedinger's Cat and Wigner's Friend[1]

---

[1] Einstein's mouse refers to Einstein's argument that since, according to quantum theory, the observer creates or partially creates the observation, a mouse can remake the universe by looking at it. Since this appears absurd, Einstein concluded that quantum physics contains some huge undiscovered fallacy. Schroedinger's Cat refers to Schroedinger's proof that a cat can exist in a mathematical condition or *eigenstate* where calling it dead and calling it alive both make sense and calling it both dead and alive also makes sense. Wigner's Friend refers to Wigner's addendum to Schroedinger, showing that even when

appear in any account of how you identify something across the room as a sofa and not as a hippopotamus. I will demonstrate and elucidate as we proceed. Meanwhile, at a reference point at the beginning, consider this:

Physicists agree that we cannot find "absolute truth" in the quantum realm but must remain satisfied with probabilities or "statistical truths". Transactional Psychology, the psychology of perception, also says we cannot find absolute truth in its field of study (sense data) and recognizes only probabilities or (some say frankly) *"gambles"*. The physicist states that in many cases we cannot meaningfully call Schroedinger's cat "a dead cat" but only "probably dead" and the Transactional psychologist says that in many cases we cannot call the Thing in the Corner a chair but "probably a chair." The simple either/or judgment — "dead" or "alive," "a chair" or "not a chair" — has become, not the only case in logic, but the extreme or limiting case, and some say only a theoretical case.

(If you feel confused, don't worry. We will examine these problems in greater detail later, and you will feel more confused.)

In short, when modern neuroscience describes how our brains actually operate it perforce invokes the same sort of paradoxes and/or the same statistical or multi-valued logic that we find in the quantum realm. Thus I dare to write about a field not my own because, in many discussions with quantum physicists, I have found the subject entirely isomorphic to my own specialty, the study of how perceptions and ideas get into our brains.

To the Transactional psychologist, quantum mechanics has the same fascination (and the same resemblance to brain science) as cryptozoology, lepufology and Disinformation Systems, and all these fields, the scientifically sober and the disreputably weird, bear a distinct family resemblance to each other.

---

the cat has become definitely alive or definitely dead for one physicist, it remains both dead and alive for another physicist located elsewhere (e.g., outside the laboratory.)

Perhaps I had better explain that. Cryptozoology deals with (a) animals whose existence remains neither proven nor disproven (e.g., the giant serpents allegedly dwelling in Loch Ness, Lake Champlain etc.; Bigfoot; the Abominable Snowman of the Himalayas etc.) and (b) animals reported in places where we don't expect them (the mountain lion of Surrey, England, the kangaroos of Chicago, the alligators in New York's sewers etc.) Those who "know" how to judge such data have not kept up with neuroscience; those who know the most about neuroscience display the greatest agnosticism about these critters and also have the greatest unwillingness to judge them.

Lepufology concerns UFO sightings in which rabbits play a significant — and usually highly puzzling — role. (Some sample cases in both cryptozoology and lepufology appear in my book, *The New Inquisition*, Falcon Press, 1987.) Again, those who "know" that lepufology cannot yield useful data usually do not know neuroscience at all, at all. Cases in which farmers claim that UFOs stole their rabbits make an ideal arena in which to test Transactional Quantum Psychology against the premature certitudes of Dogmatic Believers and Dogmatic Deniers.

Disinformation Systems consist of elaborate deceptions, constructed by intelligence agencies like the C.I.A., K.G.B. or England's M.I.5, in which a cover story, when created, has within it a second deception, disguised to look like "the hidden truth" to any suspicious rival who successfully digs below the surface. Since Disinformation Systems have multiplied like bacteria in our increasingly clandestine world, any perception psychologist who looks into modern politics will recognize that quantum logic, probability theory and strong doses of zeteticism make the best tools to employ in estimating if the President has just told us another whooping big lie or has just uttered the truth for once.

After all, even those who create Disinformation Systems have themselves swallowed Disinformation Systems devised by their rivals. As Henry Kissinger once said,

"Anybody in Washington who isn't paranoid must be crazy."

In dealing with cryptozoology, lepufology, Disinformation Systems and Quantum Mechanics one eventually feels that one has come close to total nonsense, a basic defect in the human mind (or the Universe?) or some mental fugue similar to schizophrenia or solipsism. However, as our opening drawing showed and we will see again and again, the ordinary perceptions of ordinary people contain just as much "weirdness" and mystery as all these Occult Sciences put together.

Thus I will try to show that the laws of the sub-atomic world and the laws of the human "mind" (or nervous system) parallel each other precisely, exquisitely, and elegantly, down to minute details. The student of human perception, and of how inference derives from perception, will find no shocks in the allegedly mind-boggling area of quantum theory. We live amid quantum uncertainty all our lives, but we usually manage to ignore this; the Transactional psychologist has found herself or himself forced to confront it squarely.

This parallelism between physics and psychology should occasion no great surprise. The human nervous system, after all — the "mind" in pre-scientific language — created modern science, including physics and quantum mathematics. *One should expect to find the genius, and the defects, of the human mind in its creations,* as one always finds the autobiography of the artist in the art-work.

Consider this simple parallelism: a husband and wife come to a marriage counselor seeking help. He tells one story about their problems. She tells quite a different story. The counselor, if well-trained and sophisticated, does not believe either party completely. Elsewhere in the same city, two physics students repeat two famous experiments. The first experiment seems to indicate that light travels in waves. The second seems to indicate that light travels in discrete particles. The students, if well-trained and sophisticated, do not believe either result. The psychologist, you see, knows that each nervous system creates its own model of the world, and the physics students of today know that each instrument also creates its own model of the world. Both in psychology and in physics we have outgrown medieval Aristotelian notions of "objective reality" and entered a non-Aristotelian realm, although in both fields we still remain unsure (and quick to quarrel with each other) about what new paradigm will replace the Aristotelian true/false paradigm of past centuries.

Claude Shannon's famous equation for the information content of a message, H, reads

$$H = -\Sigma p_i \log_e p_i$$

The reader terrorized by mathematics (persuaded by incompetent teachers that "I can't understand that stuff") need not panic. Σ merely means "the sum of." The symbol, p_i, tells us what we will summarize, namely the various probabilities (p_1, p_2...etc. to p_n, where **n** equals the number of signals in the message) that we can *predict in advance* what will come next. The logarithmic function merely shows that this relationship does not accumulate additively but logarithmically. *Notice the minus sign.* The information in a message equals the negative of the probabilities that you can predict what will come next every step of the way. The easier you can predict a message, the less information the message contains.

Norbert Weiner once simplified the meaning of this equation by saying that great poetry contains more information than political speeches. You never know what will come next in a truly creative poem, but in a George Bush speech you not only know what will come next, you probably could predict the whole speech, in general, before he even opened his mouth.

An Orson Welles film has more information than an ordinary film because Orson never directed a scene quite the way any other director would do it.

Since information increases logarithmically, not additively, the rate of information flow has steadily increased since the dawn of history. To quote some statistics from the French economist George Anderla (rather familiar, by now, to readers of my books) information doubled in the 1500 years between Jesus and Leonardo, doubled again in the 250 years from Leonardo to Bach's death, doubled again by the opening of our century, etc. and doubled in the seven years between 1967 and 1973. Dr. Jacques Vallee recently estimated that information currently doubles every 18 months.

Obviously, the faster we process information, the more rich and complex our models or glosses — our reality-tunnels — will become.

Resistance to new information, however, has a strong neurological foundation in all animals, as indicated by studies of imprinting and conditioning. Most animals, including most domesticated primates (humans) show a truly staggering ability to "ignore" certain kinds of information — that which does not "fit" their imprinted/conditioned reality-tunnel. We generally call this "conservatism" or "stupidity", but it appears in all parts of the political spectrum, and in learned societies as well as in the Ku Klux Klan.

To the Transactional psychologist, then, and even more to the Quantum Psychologist, something as absurd as lepufology contains many clues to how humans will, and will not, process new information.

For instance in *Flying Saucer Review*, November 1978, p. 17, one finds a report of a UFO which stole all the rabbits from a farmer's hutch. .

True or false or whatever, this report contains high information, because most of us have not heard of UFOs stealing rabbits. The signal has high unpredictability.

UFO Phenomena and B.S. edited by Haines p. 83: a close encounter in which the UFO "pilot" looked like a giant rabbit.

The information content has quantum jumped. *Two* UFO/rabbit stories?

But the Mutual Easter Bunny Observation Network, MEBON (a splinter off the less bizarre Mutual UFO Network, or MUFON) has *dozens* of these stories in their files. (They also have, as you might guess, a weird sense of humor.)

Take this as delightful whimsy or sinister nonsense, file it as you will according to your own reality-tunnel, but —

our information bank has grown richer. Dozens of UFO/ rabbit stories indicate something about UFOs or something about human psychology, something we never suspected before.

If the reader has a statistically normal reaction to this data, then she or he may understand better how the groups you dislike manage to "ignore" or otherwise resist information that seems very, very important to you...

Exercizes

1. Let every member of the group make a drawing of the room in which they meet, *as it looks from the place where they are sitting.* (This does not constitute an art contest, so don't worry if you can't draw as well as somebody else in the group.) Compare the drawings, not as "art", but as reality-tunnels. Does any one drawing seem more "true" than all the others?

2. Let every member of the group make an Architect's Drawing (i.e., a floor plan) of the room. Why do these drawings, when finished, look more alike than the drawings from individual perspectives? Discuss.

Which would you consider more "real" — the abstract Architect's floor plan — which shows something nobody ever sees in experience but which all can agree serves a useful function — or the various drawings from individual perspectives, which show the plural "realities" that people actually see, but which have no practical function?

3. Oscar Wilde said, "All art is quite useless." Discuss.

FOUR

Our "Selves" & Our "Universes"

To state our major thesis again in different words, Uncertainty, Indeterminacy and Relativity appear in modern science for the same reason they appear in modern logic, modern art, modern literature, modern philosophy and even modern theology. In this century, *the human nervous system has discovered its own creativity, and its own limitations.*

In Logic, for instance, we now recognize not only "meaningless" propositions but also "Strange Loops" (systems containing concealed self-contradiction) both of which can infest any logical system, like a virus invading a computer — but these logical "bugs" have often lingered for centuries before being discovered.

People have murdered each other, in massive wars and guerrilla actions, for many centuries, and still murder each other in the present, over Ideologies and Religions which, stated as propositions, appear neither true nor false to modern logicians — meaningless propositions that look meaningful to the linguistically naive. (For instance, much of this book will attempt to show that every sentence containing the innocent-looking word "is" also contains a hidden fallacy. This will come as a distinct shock, or will seem like Crazy Heresy, to those Americans currently battling in rival "demonstrations" and acts of civil disobedience over the question of whether a fetus — or even a zygote — "is" or "is not" a human being.)

Meanwhile, in Art, Picasso and his successors have shown us that a work of, say, sculpture can move us deeply even if it has opposite meanings like our two-face drawing. One Picasso classic moves me, for instance, even though I can see it either as the head of a bull or the seat and handlebars of a bicycle.

Joyce's *Ulysses* mutated the novel by describing one ordinary day, not as an "objective reality" in the Aristotelian sense but as a labyrinth in which nearly a hundred narrators (or "narrative voices") all report different versions of what happened. Different reality-tunnels.

Modern philosophy and modern theology have arrived at such resonant conclusions as "There are no facts, only interpretations" (Nietzsche) or "There is no God and Mary is His mother" (Santayana) or even "God is a symbol of God." (Tillich)

All this results from our new awareness of our "selves" as the co-authors of our "universes". As Dr. Roger Jones says in *Physics as Metaphor*, "whatever it is we are describing, the human mind cannot be parted from it." Whatever we look at, we must see, first and foremost, our own "mental filing cabinet" — the structure of the software which our brain uses to process and classify impressions.

By "software" I mean to include our language, our linguistic habits, and our over-all tribal or cultural world-view — our game-rules or *unconscious* prejudices — the tacit reality-tunnel which itself consists of linguistic constructs and other symbols.

In daily life, the software of most readers of this book consists of Indo-European language categories and Indo-European grammar. In advanced science, the software includes both of these and also the categories and structures of mathematics, but in either kitchen-sink problems or nuclear reactor problems we "see" through a symbolic or semantic grid, since math, like language, functions as a code *which imposes its own structure on the data it describes.*

The painter "thinks" (when painting) in form and color, the musician in sound frequencies, etc. but most human mentation employs words most of the time, and even specialists like the mathematician, painter, musician etc. use words for a large part of their thinking.

Whatever we know, or think we know, about our "selves" and our "universes", we cannot communicate about either inner or outer realms without using language or symbolism — brain software. To understand this book, the reader must remind herself (or himself) again and again, that even in thinking, and even in special areas like math and art, we use some kind of symbols to "talk to ourselves" or visualize.

The only "thing" (or process) precisely equal to the universe remains the universe itself. Every description, or model, or theory, or art-work, or map, or reality-tunnel, or gloss, etc. remains somewhat smaller than the universe and hence includes less than the universe.

What is left in our sensory continuum when we are neither talking nor thinking remains non-symbolic, non-verbal, non-mathematical — ineffable, as the mystics say. One can speak poetically of that non-verbal mode of apprehension as Chaos, like Nietzsche, or the Void, like Buddha; but "Chaos" and "the Void" remain only words and the experience itself stubbornly remains non-verbal.

At that point one can only correctly say, with Wittgenstein in his *Tractatus Logico Philosphicus*: "Whereof one cannot speak, thereof one must remain silent." Zen Masters merely point or wave their staff in the air.

When we leave the nonverbal, when we again talk or think, we perforce make symbolic maps or models, which cannot, by definition, equal in all respects the space-time events that they represent. This seems so obvious that we all paradoxically never think about it and hence tend to forget it. Nonetheless, a menu does not taste like a meal, a map of New York does not smell like New York (thank God), and a painting of a ship in stormy waters does not contain the captain and crew who have to deal with real ships in real storms.

All kinds of maps or models also show, on examination, the personality or "mental furniture" of their creator, and, to a lesser extent, of the creator's society and linguistic system(s) — the semantic environment.

An experienced sailor will quickly spot the difference between a painting of a ship by somebody who has also worked as a sailor and a very similar painting by somebody who has only read about sailing.

Many a novel or play written in 1930, which seemed "brutally realistic" then, now seems a little quaint and "unreal" in places, because we no longer live in the semantic environment of 60 years ago. Joyce's *Ulysses* escaped this trap by not having a point of view at all, at all — his multiple narrator technique gives multiple points of view — just as post-Copenhagen physicists escape it by what they call *"model agnosticism"*, not accepting any one model as equal to the whole universe.

Consider a map that tries to show, not "all" the universe, but something less ambitious — all of Dublin, Ireland. Obviously, the map would have to occupy the same amount of space as Dublin. It would also have to include about a trillion moving parts at least — one and half million humans, an equal number of rats, a few million mice, perhaps billions of bugs, hundreds of billions of microbes, etc.

To tell "all" about Dublin this map would have to let its moving parts go on moving for at least 2000 years, since a town (not always called Dublin) has stood on the river Anna Liffey for about that long.

This map would still not tell "all" about Dublin, even up to this date (excluding the future...) until it somehow included all the thoughts and feelings of the human and other inhabitants of that area...

At this point, the map would still prove mostly useless and largely irrelevant to a geologist, who wants to know the chemistry and evolution of the rock and soil on which Dublin stands.

So much for the "external" world. What kind of map would ever approximate toward telling "all" about *you?*

Exercizes

1. Let a sexual partner team (husband/wife or two lovers) re-enact their most recent quarrel. (If nobody will admit that they "quarrel", let the chosen subjects re-enact their most recent disagreement.)

2. Let this couple then reverse roles and let each one "play" the other in a continuation of the disagreement. Attempt to employ the technique of Method Acting: let each player try to *feel* the point of view of the other while acting the other.

3. See if you have two people in the group with opposed views on some "hot" issue (e.g., abortion, gun control, the war on drugs, etc.) Let them each attempt, by Method Acting, to present the point of view of the other, as sincerely as possible.

4. Let one member of the group acquire the following thirteen items:

a toy fire-truck;
a Barbie doll;
a reproduction of a Picasso painting.
a brick;
a screw-driver;
a hammer;
a turkey feather;
a piece of balsa wood;
a rubber ball;
a piece of hard wood, such as birch;
a "ghetto blaster" (portable stereo);
a pornographic novel;
a philosophical treatise by Bishop George Berkeley.

Place these items on the floor and let everybody sit around them. First, divide them into two groups — red things and not-red things. See how many times ambiguous cases arise (e.g., should a book with a red-and-white cover go in the red pile or the not-red pile?)

Let the 13 items be divided into another two groups — useful objects and toys. See how many ambiguities arise. (Does art belong among toys? Does pornography?)

Each week, as long as the group continues, let somebody think of another dualism and divide the 13 items into two piles according to that new dichotomy.

Note each case where two things fall into different groups according to one dualist system fall into the same group accord-

ing to another dualist system. (E.g., balsa wood and hard wood will fall into the same group if one divides "wooden things" from "non-wooden things," but will fall into different groups if one divides "things that float" from "things that do no not float.")

Note how the Aristotelian argument "It 'is' either an A or a not-A" appears after you have found several things that belong on the same side of one dualism but on opposite sides of other dualisms.

Some suggestions for other dualisms: "educational things" and "entertaining things," "scientific things" and "non-scientific things," "good things" and "bad things," "organic things" and "inorganic things."

See how many odd and imaginative dualisms the group can create.

At this point, an obvious fact seems worthy of special emphasis. *Actually doing these exercizes in a group, as suggested, teaches much more than merely reading about them.*

FIVE

How Many Heads Do You Have?

Borrowing a joke (or a profundity?) from Bertrand Russell's *Our Knowledge of the External World*, I will now demonstrate that the reader has two heads.

According to common sense, and the consensus of most (Occidental) philosophers, we exist "inside" an "objective universe", or — to say it otherwise — the "objective universe" exists "outside" us.

Very few people have ever doubted this. Those who have doubted it have arrived, inevitably, at highly eccentric conclusions.

Well, then, avoiding eccentricity and accepting the conventional view, how do we know anything about that "external universe"? How do we perceive it?

(For convenience, I will consider only the sense of sight in what follows. The reader can check for himself, or herself, that the same logic applies if one changes the terms and substitutes hearing or any of our other senses.)

We see objects in the "external universe" through our eyes and then make pictures — models — of them in our brains. The brain "interprets" what the eyes transmit as energy signals. (For now, we will ignore the data that shows that the brain *makes a gamble* that it can interpret these signals.)

Again, very few Occidentals have doubted this, and those who have doubted it all arrived at strange and incredible alternatives.

So, then, we live "inside" an "external universe" and make a picture or model of it "inside" our brains, by adding together, or synthesizing, and interpreting, our pictures or models of parts of the universe called "objects". Then, it follows that we never know the "external universe" and its "objects" at all. *We know the model of the "external universe" inside our brains, which exist inside our heads.*

In that case, everything we see, which we think of as existing externally, actually exists internally, inside our heads.

But we have not arrived at solipsism, remember. We still assume the "external universe" from which we started. We have merely discovered that we cannot see it or know it. We see a model of it inside our heads, and in daily life forget this and act as if the model exists outside our heads — i.e., as if (1) the model and the universe occupy the same area of space (as our map that tries to show "all" about Dublin would occupy the same space as Dublin) and (2) that this space exists "outside".

But the model and the universe do not occupy the same space and the space where the model exists can only be located "inside" our brains, which exist inside our heads.

We now realize that, while the universe exists outside, the model exists inside, and therefore occupies much, much less space than the universe.

The "real universe" then exists "outside" but remains unexperienced, perhaps unknown. That which we do experience and know (or think we know) exists in local networks of electrochemical bonds in our brains.

Again, if the reader cares to challenge any part of this, she or he should certainly try to imagine an alternative explanation of perception. It will appear, or it has always appeared to date, that any and all such alternatives sound not only queerer than this but totally unbelievable to "people of common sense."

Well, to proceed, we have now an "external universe", very large (comparatively speaking) and a model of same, much smaller (comparatively speaking), the former "outside" us and the latter "inside" us. Of course, some correspondence or isomorphism exists between the "external" and "internal" universes. Otherwise, I could not get up from my chair, walk to the door, go down the hall and accurately locate the kitchen to get another cup of coffee from something I identify as a Coffee Maker.

But where does our head exist?

Well, our head obviously exists "inside" the "external universe" and "outside" our brain which contains the model of the "external universe".

But since we never see or experience the "external universe" directly, and only see our model of it, we only perceive our head as *part of the model*, which exists inside us. Certainly, our perceived head cannot exist apart from our perceived body as long as we remain alive, and our perceived body (including head) exists inside our perceived universe. Right?

Thus, the head we perceive exists *inside* some other head we do not, and cannot, perceive. The second head contains our model of the universe, our model of this galaxy, our model of this solar system, our model of Earth, our model of this continent, our model of this city, our model of our home, our model of ourselves *and atop our model of ourselves a model of our head. The model of our head thus occupies much less space than our "real" head.*

Think about it. Retire to your study, unplug the phone, lock the door and carefully examine each step of this argument in succession, noting what absurdities appear if you question any individual step and try an alternative.

Let us, for Jesus sake and for all our sakes, at least attempt to clarify how we can have two heads. Our *perceived* head exists as part (a very small part) of our model of the universe, which exists inside our brain. We have already proven that, have we not? Our brain, however, exists inside our second head — our "real" head, which contains our whole model of the universe, including our perceived head. In short, our perceived head exists *inside* our perceived universe which exists *inside* our real head which exists *inside* the real universe.

Thus, we can name our two heads — we have a "real" head outside the perceived universe and a "perceived head" inside the perceived universe, and our "real" head now appears, not only much bigger than our perceived head, but *bigger than our perceived universe.*

And, since we cannot know or perceive the "real" universe directly, our "real" head appears bigger than the only universe we do know and perceive — our perceived universe, inside our perceived head.

The reader might find some comfort in the thought that Bertrand Russell, who devised this argument, also invented the mathematical class of all classes that "do not contain themselves." That class, you will note, does not contain itself unless it does contain itself. Also, it does contain itself if and only if it does not contain itself. Got it?

When not busy crusading for rationalism, world peace, common decency, and other subversive ideas, Russell spent a lot of time in the even more subversive practice of inventing such logical "monsters" to bedevil logicians and mathematicians.

Returning to our two heads: Lord Russell never carried this joke, or this profound insight, beyond that point. With a little thought, however, the reader will easily see that, having analyzed the matter this far, we now have *three* heads — the third containing the model that contains the "real" universe and the "real" head and the perceived universe and the perceived head. And now that we have thought of that, we have a fourth head...

And so on, *ad infinitum.* To account for our perception of our perception — our ability to perceive that we perceive — we have three heads, and to account for that, four heads, and to account for our ability to carry this analysis onward forever, we have infinite heads...

A model of consciousness which does arrive, very rigorously and with almost mathematical precision of logic, at precisely this infinite regress appears in *The Serial Universe* by J. W. Dunne, who uses time instead of perception as his first term but still arrives at the conclusion that we have, if not an infinite series of heads, an infinite series of "minds".

Like the Zen teacher, I have just led you to the door of the Law and slammed it in your face. But wait. We will eventually discern "light at the end of the tunnel." If we can only open that damned door...

Or perhaps you have detected Mr. Fields' "Ubangi in the fuel supply" already?

If not, let us proceed. Alfred Korzybski, mentioned here several times (and a strong influence even when not mentioned), urged that our thinking could become more scientific if we used mathematical subscripts more often.

Thinking about this one day, I came up with the following analog of Dunne's argument without even using his infinite time dimensions:

I observe that I have a mind. Following Korzybski, let us call this observed mind, $mind_1$.

But I observe that I have a mind that can observe $mind_1$. Let us call this self-observing mind, $mind_2$.

$Mind_2$ which observes $mind_1$ can in turn become the object of observation. (A little experience in Buddhist self-observation will confirm this experimentally.) The observer of $mind_2$ then requires its own name, so we will call it $mind_3$.

And so on...to infinity, once again.

Of course, having mentioned Buddhism, I might in fairness add that the Buddhist would not accept "I observe that I have a mind." The Buddhist would say "I observe that I have a tendency to posit a mind."

But that, perhaps, allows the *felix domesticus* to escape the gunnysack, as Mr. Fields would say.

Exercizes

1. Let the group look back at Exercize 1 at the end of Chapter Two. Try to decide how many of the propositions there, which I then asked you to force into the two categories "meaningful" and "meaningless" might fit just as well into the category of Game Rules *or the resultants of tacit (unstated) Game Rules.*

2. Meditate upon the following quote from Lord Russell's *Our Knowledge of the External World* (page 24):

The belief or unconscious conviction that all propositions are of some subject-predicate form — in other words, that every fact consists of some thing having some quality — has rendered most philosophers incapable of giving any account of the world of science and daily life.

Consider the subject-predicate form as a Game Rule.

3. Contemplate the following typical subject-predicate sentences: "The lightning flashed suddenly." "It is now raining out." "I have an uncontrollable temper."

Try to identify the subject, "it" in the sentence: "It is now raining out."

See how subject-predicate Game Rules influence the other two sentences. Can any of you restate them in more phenomenological language?

Does any of this help you see the trick in the two-heads (or infinite heads) argument?

SIX

The Flight From Reason &
The Cult of Instruments

Long, long before modern physics or modern psychology, in ancient Greece, the Skeptics had already noticed that Uncertainty, Indeterminacy and Relativity appear inescapable parts of human life, because what Xerox sees is never exactly what Exxon sees. Plato, Aristotle and other geniuses attempted to escape the agnosticism or Zeteticism of the Skeptics by "finding," or claiming to find, a method of Pure Abstract Reasoning that, they believed, would arrive at Pure Truth without any distortions introduced by our fallible human sense organs. Aside from a few conservatives in Chairs of Philosophy, the world now realizes that the Greek search for such Pure Truth failed; and the subsequent history of philosophy seems like a long detective story — the gradual discovery, century after century, of the numerous "lies" (unconscious prejudices) that crept into the Pure Reasoning of those bold Hellenic pioneers.

To speak caustically about it, one might say the Greek Logicians suffered from the illusion that *the universe consists of words.* If you found the right words, they seemed to think, you would have Eternal Truth.

Then came modern science, a synthesis of Pure Reason in the Greek tradition with humble empiricism in the tradition of the craftsmen and artisans — with all results expressed in the very precise special "languages" of various branches of mathematics. It seemed, for a few centuries, that science could solve all mysteries and answer all questions. In science, reasoning about what the

Universe "should do" (according to Logic) lived in a marriage or feedback loop with increasingly subtle instruments to tell us where and when the Universe failed to agree with our Logic or our math — where our Logic needed revision, or one type of math needed correction by another. With sufficiently perfected instruments, it then seemed, we could correct all our errors and arrive eventually at the Pure Truth which Plato and Company had thought they could trap with mere Logic itself, without the instruments.

The universe now seemed to consist, not of words, but of equations. Someday, we thought, we would know "bloody all about bleeding everything," and describe it in elegant mathematical formalisms. That faith died with Einstein's Relativity and Plank's Quantum Mechanics, both of which discovered, in different ways, that the human nervous system aided by humanly designed instruments produces results no more "infallible" than the human nervous system unaided by instruments.

As an illustration: the Skeptics in ancient Greece had observed the relativity of temperature as perceived by humans. Every philosopher in Athens had heard their experimental argument: if you hold your right hand in a bowl of rather hot water and your left in a bowl of very cold water, and then dunk both hands in a third bowl of tepid water, the right hand will "read" the third bowl as cold and the left will "read" it as hot.

The whole heroic effort of Plato and Aristotle, as we said, amounted to an attempt to get beyond this sensory-sensual relativity by use of Pure Reason.

Pure Reason, however, derives from axioms which can neither be proven nor disproven. These axioms appear in consciousness from a level of pre-logical apprehension in which we might as well be gesticulating and pointing — or waving sticks in the air like Zen Masters — instead of talking, because we are trying to indicate or invoke something that exists before words and categories.

Worse: the axioms (game rules) that seem natural or undeniable in one tribe or culture do not seem at all natural and are often denied in other cultures. Hence, most of the "self-evident" axioms of Plato and his associates no longer win assent from scientists, and many of them have turned out to disagree with actuality (nonverbal experience) when scientists tried to check them.

Immanuel Kant perhaps composed the longest list of defects in classical Greek "pure reason". One that has received less publicity that most — much less publicity than the Cretan who says Cretans always lie — goes like this:

When an arrow gets fired from a bow toward a target it appears to move through space.

However, at every instant the arrow actually occupies one position in space, not two or three or more positions.

Thus, at every instant the arrow exists in one place, not in two or three or more. In other words, at every instant the arrow has a position.

If the arrow has one and only one definite position at every instant, then at every instant it does not move. If it does not move at any of these instants, it never moves at all.

You cannot escape this Logic by positing instants-between-instants. In these nanotime units, the same logic holds. At each nano-instant, the arrow has some location, not several locations. Therefore, even in nano-instants, the arrow does not move at all.

It seems the only way out of this absurdity consists of claiming that the arrow does, after all, occupy two locations at the same time. Alas, this leads to worse problems, which I leave you to discover for yourself.

And that shows where Logic gets you, if uncorrected by observation (senses or instruments). If we do not correct our Logic by comparing it with experience, we may go on for centuries elaborating our most ancient errors endlessly — as seems obviously to have happened to cultures that do not share our "self-evident" axioms.

But we seem as nutty to those cultures as they seem to us. Every religion, for instance, seems to other religions (and nonbelievers) the result of logical deductions from axioms that just don't fit this universe.

So, then, let us by all means correct our Pure Reason with actual experience of what people see and smell and otherwise detect in the phenomenological or existential world. Let us expand beyond abstract Pure Reason and check our logic against our experience.

So, then: from this kind of argument, science emerged — and seemed for a while ready and able to solve all our problems.

Certainly, with its splendid equations and marvelous instruments, science seemed to offer a better way to solve nitty-gritty existential problems than Greek logic ever had. Businessmen noticed this quickly and began funding "research". Rationalist philosophers noted it later and joyously assumed science could go beyond practicality — the best model at a date — and also produce Pure Truth.

But then Einstein showed that two clocks can measure different times — just like the two human hands "measuring" different temperatures. The fallibility of our nervous systems suddenly appeared also in our instruments; and Absolute Truth again eluded us.

Einstein, to repeat for emphasis, also demonstrated that two rulers can measure different lengths. Then Quantum Mechanics showed that different instruments can yield radically different "readings" of space-time events in the sub-atomic world. In the most shocking case of all, one which still sends first year physics students reeling, one instrumental set-up shows us a world made of discrete bullet-like particles and the same instruments in a different set-up show a world made of ocean-like energy waves.

This seemed "incomprehensible" to physicists at first because, three hundred years after Galileo shot Aristotle's physics full of holes, they were still thinking in the categories of Aristotle's logic, where X must "be" either a wave or a particle and can't possibly "be" both a wave and a particle, depending on how and when we "look" at it. For a while, some physicists were actually talking, facetiously, but also a bit desperately, about "wavicles".

In summary: we thought we could escape the relativity and uncertainty of sense organs by building smart instruments, but now we have discovered the relativity of the instruments themselves. (I keep reiterating this because, in my experience teaching seminars on non-Aristotelian logic for 30 years, hardly anybody understand this at first. Most people think they understand it, but they don't.)

Thus, when you examine a rose bush, whether you look with your eyes (and brain) alone, or look with a variety of scientific instruments, what you will "see" depends on the structure of the instrument — your sensory apparatus and/or the tools added to that apparatus.

Furthermore, what you can say about what you saw depends on the structure of your symbolism — whether you describe it in English, Persian, Chinese, Euclidean geometry, non-Euclidean geometry, differential calculus or quaternions.

This explains why, in Dr. Jones's words, "whatever we are describing, the human mind cannot be parted from it."

Exercizes

1. Weather permitting, leave the house, go outside to the street and look around. How much of what you see would have existed if humans had not designed and built it? How much that "just grew there" would look different if humans had not cultivated and encouraged (or polluted) it?

2. Look at the sky. If you can distinguish stars from planets, can identify some of them, etc., try to forget this knowledge and imagine how the sky looks to very intelligent animals without human science. Then look at it again with your knowledge of astronomy back in focus.

3. If a meteor passes, how does it make you feel when trying to see without scientific glosses? How differently do you feel when you allow yourself to remember what you know of meteors?

4. Go back inside and discuss this:

If all TV shows about the police (about 20 a week in most areas) went off the air and instead we had an equal number of TV shows about landlords, would this change the average American reality-tunnel?

In how many ways would the reality-tunnel change?

What would Americans "see" (or remember) that they now tend to ignore? What would they become less aware of? What would they become much more aware of?

5. Try to figure out why there are so many TV shows about police and virtually no shows about landlords.

Who decides this? Why have they decided it this way? (Attempt to avoid paranoid speculations or grandiose conspiracy theories, if at all possible.)

SEVEN

Strange Loops & the Infinite Regress

If we never describe anything "as it is" but only "as it appears to our minds," we can never have a pure physics, but only neuro-physics — i.e., physics as known through the human nervous system. We can also never have pure philosophy, but only neuro-philosophy — philosophy as known through the human nervous system. And we can never have pure neurology but only neuro-neurology — neurology as known through the human nervous system...

But at this point we have already entered the arena of Strange Loops, as some readers have guessed, for neuro-neurology can only be known by the nervous system and thus can only be known by a meta-science of neuro-neuro-neurology...which can only be known through neuro-neuro-neuro-neurology...and so on, *ad infinitum.* Do you detect Lord Russell's two-head argument looming on the horizon at this point? Or even J. W. Dunne's infinite regress of consciousnesses in time?

Some Zen bastard seems to have slammed the door of the Truth in our face again.

This neurological regress precisely parallels a proof in Quantum Mechanics, known as "Von Neumann's Catastrophe" (or "Von Neumann's Catastrophe of the Infinite Regress", in full) which shows that we can add an infinite number of instruments to our existing instruments and still never get rid of some degree of Uncertainty and Indeterminacy. (By the end of this book the reader will hopefully understand why this "coincidence" and dozens like it inescapably link Quantum Mechanics with daily-life psychology or ordinary kitchen-sink consciousness.)

At this point some readers may want to bail out or throw the book away, thinking I shall soon lead them into the bottomless abyss of solipsism or some neo-Berkeleyan Idealism. Not at all: a stark dualism of Certainty versus Uncertainty only appears in two-valued Aristotelian logic. In mathematical logic, we do not have to choose between Pure Certainty and Pure Uncertainty. Rather, in the mathematics of probability, we have infinite choices in between those extremes.

For convenience we can reduce these to the traditional 100 used in ordinary percentages.

Thus, if Pure Certainty equals 100% and Pure Uncertainty equals 0%, the logic of Quantum Mechanics and of the Quantum Psychology in this book does not tell us that the impossibility of reaching 100% leaves us stuck at 0% forever. Quite the reverse. Many things in daily life have probabilities over 50%, which will satisfy any gambler and keep up his interest; even better, some things have probabilities of 90%, 95% or even higher.

Personally, I never worry about the thermodynamic fact that the probability of air remaining approximately evenly distributed around this room never reaches 100%. The probability that all the air will suddenly rush to one corner and leave me to die in a vacuum has been calculated as greater than 0% and much, much less than 0.001%, but I refuse to get anxious about it.

The probability that I will get hit by a meteor tomorrow seems much, much higher — maybe almost as high as 0.1% — but I don't worry much about that either.

The business person, like the physicist or gambler, has long grown accustomed to this aspect of Quantum Psychology. Businesses do not hope for 100% certainty in making decisions (i.e., they do not judge grain futures by religious Dogma) but they do not muddle about in endless Hamlet-like indecision (total uncertainty), either. They long ago learned to "guesstimate" or intuit probabilities, and nowadays they have generally graduated from guesstimating to precise estimating with mathematical probability matrices in a computer.

Thus, the "loss of certainty" does not mean a descent into the void of solipsism. It merely means a graduation from the kindergarten level of "yes" (100%) or "no" (0%) to the adult world of "how closely can we calculate the odds on this happening?"(5%? 25%? 75%? 95%?)

I must admit, however, that the logic of probability does lead to some weird implications. In this connection, consider what I call the Jesus H. Christ Night.

Most math students, early on in their University years, encounter the paradox of the Paddy Murphy Night. The odds of getting two Paddy Murphys in the same class seems small, but it does happen. What seems distinctly odd to non-mathematicians — "the Paddy Murphy" paradox — consists in this: if the universe lasts long enough, some lecturer must eventually confront a class consisting entirely of men named Paddy Murphy. If you think about it, you will easily see, intuitively, that this Paddy Murphy Night must eventually occur. What staggers most people lies in the result we get if we imagine a universe that last an infinite number of years.

In that infinite universe, Paddy Murphy Night not only occurs once, or several times, but an infinite number of times. (However, non-Paddy Murphy Night also occurs an infinite number of times. This illustrates Cantor's principle that if you remove an infinite set from an infinite set, another infinite set remains...)[1]

Yesterday (2/3/1990) I heard a popular talk-show host (Dick Whittington, KABC, Los Angeles) mention that, in high school, in The Bronx, New York, he actually had a classmate named Jesus Christ.[2] Mr. Whittington returned

---

[1] For example, the set of whole numbers continues to infinity, but so does the set of even numbers. If you subtract the even numbers from the whole numbers, you still have an infinite set of odd numbers.

[2] Mr. Whittington remembered this because of a news bulletin concerning a man named Joe Blow, who complained that his name created problems in job-hunting. People would start laugh-

to this topic a few times, seemingly worried that his audience suspected him of a put-on. I felt inclined to believe him, because when I attended high school, in Brooklyn, I had a class-mate named Sven Christ, who told me that the Scandinavian countries had many families named Christ. Since many Hispanic families name their first son Jesus, which they pronounce *Hay-zeus* but most non-Hispanic Americans pronounce *Gee-zuz*, a Scandinavian-Hispanic marriage could easily produce a son named Jesus Christ.

But then I remembered Paddy Murphy night, and realized that if the universe lasts long enough, some lecturer will eventually confront an audience made up entirely of men named Jesus Christ. And in an infinite universe, this will happen an infinite number of times.

And, since Harry remains a popular middle name, some lecturer, even in a finite universe, *might* some night confront an audience consisting of men named Jesus H. Christ (or a mixed audience of Mary Christs and Jesus H. Christs). In an infinite universe, an infinite number of lecturers will encounter an infinite number of such audiences.

However, although no mathematician would dispute this, I will not live in eager anticipation of the night when I, a frequent lecturer, encounter this audience of Jesus H. Christs.

Just as I do not live in dread of all the molecules rushing to the corner and leaving me to die in a vacuum.

I emphasize and will reiterate this because so many people have been hypnotized by Aristotelian "yes/no" logic to the extent that any step beyond that Bronze Age mythos seems to them a whirling, dizzying plunge into a pit of Chaos and the Dark Night of Nihilism.

This book on Quantum Psychology, then, attempts to show that the Uncertainty and Indeterminacy of quantum physics has its origin in our brains and nervous systems; that all other knowledge has the same origin; and that the

---

ing when they saw his name on an application, Mr. Blow said, and could not seem to take him seriously as a possible employee, as if someone named Porky Pig had asked for a job.

non-aristotelian logics invented by quantum physics describe all other efforts of human beings to know and to talk about the world of experience, on any level.

Mr. A in his office trying to understand why his boss acts "unfairly," and Dr. B in her laboratory trying to understand why a quantum function behaves as it does, must both always remain part of a seamless unity with what they seek to understand.

I do not consider this book Quantum Philosophy, however. I have called the ideas herein Quantum Psychology because the consequences of Relativity, Uncertainty and Indeterminacy have literally earth-shaking implications for our daily lives, our "mental health," our relations with other humans, and even our deepest social problems and our relations with the rest of the Earth and the Cosmos itself. As Count Alfred Korzybski noted in the 1930s, *if all people learned to think in the non-Aristotelian manner of quantum mechanics, the world would change so radically that most of what we call "stupidity" and even a great deal of what we consider "insanity" might disappear,* and the "intractable" problems of war, poverty and injustice would suddenly seem a great deal closer to solution.

Think about it.

The quest for Certainty in a world of Uncertainty creates some amusing parallels between the life of an individual and that of a civilization.

For instance, consider a hypothetical Joe Smith, born in Canton, Ohio, in 1942. By the time of his tenth birthday in 1952, Smith had probably arrived at premature certainty for the first time. He "believed in" various doctrines because his parents did — e.g., the superiority of the Republican Party over all others, the similar superiority of the Episcopalian Church, the desirability of racial segregation, the inevitability of all institutions (Church, State, business etc.) remaining male dominated, and the necessity of destroying World Communism, which all good people (he knew) recognized as the major Evil on the planet.

By 1962 this particular Joe Smith, then 20 years old, had arrived at Harvard and had mutated completely — taken a quantum jump. He majored in sociology, considered himself a Liberal, had severe doubts about the superiority of Republicans and Episcopalians, and thought some kind of *modus vivendi* with the Communists had to occur or the world would blow itself up. He also felt "opposed to" segregation but you would not find him doing anything practical about it, and he still hadn't questioned male superiority. He had again reached premature certainty and believed the views of the professors he liked best represented the view of "all educated people." His parents now seemed "ignorant" to him, although he felt ashamed to think that.

Joe had no idea that the Revolution of the 1960s would mutate him and his reality-tunnel in dozens of ways he could not predict in 1962. He didn't foresee Freedom Rides and Mississippi cops and clubs and tear-gas and LSD and Woodstock and the Pentagon Demonstration and Women's Liberation in his future at all, at all.

By 1972, Joe and some friends set a bomb in an unoccupied laboratory at night, to protest the use of technology in a war he considered immoral. The United States government, and not Communism, now seemed to him the supreme Evil in the world. He spouted Marxist jargon, mixed with hippie mysticism, and again, having lived 30 years, he had premature certainty.

Joe has probably spent most of the years since 1972, first living a grubby underground life while waiting for the statute of limitations to expire, and then "trying to get his head back together again" — *running on empty,* in the excellent metaphor of a recent film.

Similarly, Western Civilization reached premature certainty with Plato and/or Aristotle, reached a new kind of premature certainty with Aquinas and the medieval theologians, reached a third premature certainty with Newton and the Age of Reason etc. Today, the best educated appear as if "trying to get their heads back together again" and "running on empty." Western Civilization also has no suspicion that the Revolutions of the next two decades will mutate it and its latest reality-tunnel in dozens of ways we cannot predict in 1990...

Review Exercizes

1. Have one member of the study group find a small rock that fits easily into the human hand. At the weekly meeting, pass the rock around. Allow each person to hold and examine the rock and attempt to say "all" about it

Continue this exercize until everybody realizes that we never can say "all" about even a simple rock, or until everybody becomes embroiled in a debate between those who think eventually, in a few million years perhaps, we can say "all" and those who think we can *never* say "all."

2. Have those who think we can eventually say "all" about the rock set out to investigate the geological history of the region where the rock comes from and report the following week on "all" the history of the forces that produced the region that produced that particular rock. Have everybody else try asking questions to find important areas of information left out of this attempt to say "all."

3. Attempt the same exercize with the room in which the group meets. Have everybody take turns attempting to tell "all" about the room. Then have somebody prepare a report for next week on "all" about how the house came to have its distinct design and location and that room within it.

4. Have each person sit silently and write a description of the house in which the group meets. Take about five minutes. Read the descriptions aloud, noting on a blackboard or large pad:

   (a) how many things appear on some lists and not on others.
   (b) how many things do not appear on any lists but can quickly come to light with further investigation.

5. Have every person close her or his eyes and listen to the sounds in the room and the sounds coming in from outside. Let one person with a watch time this exercize to last two minutes, then compare reports. Note how each nervous system has heard different sounds.

6. Let the group attempt to say "all" about the city where they meet.

7. Let the group attempt to say "all" about the economic history of the city.

8. Let the group attempt to say "all" about the geological, ecological and economic history of the region in which the city exists.

9. Let the group again pass the rock around, silently. Let every person to look at it in the manner of Zen meditation — *without forming words in their heads.* (Those without experience in meditation will find this very difficult, but try it anyway.)

10. Note especially the points at which any members of the group begin to *resist* the exercizes — e.g., complain "This is silly," "I know this already," "This is some kind of put-on," etc. Note any symptoms of irritability. Pass no judgments on one another when such reactions appear. Discuss the factors that make these exercizes appear "boring" (uninteresting) or "threatening" (too interesting) to some kinds of people.

11. In another book, I suggested the new word *sombunall* meaning "some but not all." In the week after doing the above exercizes, let each member of the study group try to remember to ask, each time the word "all" occurs, "Can we safely say 'all' here? Do we know enough? Would *sombunall* perhaps fit the facts more closely?"

PART TWO

Speaking About the Unspeakable

It used to be thought that physics describes the universe.
Now we know that physics only describes what we can
say about the universe.
— Niels Bohr

"Reality? We don't got to show you no
steeeeenking reality."
— Dr. Nick Herbert, caricaturing the Copenhagen
Interpretation, Esalen Institute, February 1986

EIGHT

Quantum Logic

Dr. John von Neumann, one of the leading proponents of Bohr's view that science cannot find "one deep reality" underlying all relative instrumental realities, went one step further than Bohr. Since the quantum world just does not fit Aristotelian either/or logic, von Neumann invented a three-valued logic that suits the quantum world better.

Aristotle left us with the two choices, "true" or "false". Von Neumann added a *"maybe."* This corresponds in some ways with Dr. Rapoport's "indeterminate" state, but differs in other ways; it definitely excludes the "meaningless", which von Neumann, like Bohr, banned from scientific discourse entirely.

Some physicists (e.g., Dr. David Finkelstein) believe that von Neumann has solved "all" (or maybe sombunall?) the paradoxes that still linger even after we have accepted Bohr's Cophenhagenian rejection of "deep reality". Others regard 3-valued Quantum Logic as a mere "formalism" or "trick" and not a true contribution to clarifying the Indeterminacy and Uncertainty of quantum events.

Nonetheless, QL (quantum logic) applies very well to ordinary affairs — quite contrary to the opinion of those who assure us that quantum uncertainty does not invade our daily lives and remains only on the sub-atomic level.

For instance, I toss a coin in the air. Unless it lands on edge (a rare event) the coin definitely settles into an Aristotelian either/or when it hits the floor — heads or tails. No *maybe.*

But in what state does the coin exist while flip-flopping up from my hand and down to the floor? Some metaphysical doctrine of predestination may claim the coin exists as heads or tails even before landing, because it has been predetermined that the coin will land that way. Scientifically, such a proposition lies beyond the reach of testing, and so we must consider it "meaningless". On the operational or phenomenological level, the coin appears in a von Neumann *"maybe"* state *until* it lands.

Similarly, Transactional psychology reveals that perceptions begin always in the *"maybe"* state. I walk down the street and see good old Joe half a block away. If I haven't studied brain science, I feel sure that the Joe I see "is really" there, and I feel quite surprised when the figure comes closer and I now see a man who only resembles good old Joe slightly. My perception contained a *"maybe"* but, conditioned by Aristotelian logic, I ignored this and my conception leaped to premature certainty. (This description has been simplified for logical clarity. In experience, the feedback loop *from perception to conception and back to perception* operates so quickly, that we "see" what we think we should see and the *maybe* virtually never registers — until we re-train ourselves to register it.)

Whether something belongs in the "true/false" class or in the *maybe* class usually depends on *time* considerations. The coin belongs in *maybe* for a few seconds, while in the air, but settles into either/or when it lands. "Mary did not come to class today" will seem true to the teacher, seeing absence-of-Mary or (technically) "non-presence of Maryness" in the class, but becomes a *maybe* if somebody alleges that Mary has appeared in the distance, rushing toward class; "Mary did not come to class today" can then shift to the false category, replaced by "Mary arrived late for class," the second she enters the room.

Many perceptions not only begin in the *maybe* state but remain *maybes* forever, because the space-time events that trigger them do not last long enough for us to justify definite verdicts. Nonetheless, we ignore this and, guided by Aristotelian habit, assign definite verdicts anyway. Such seems the explanation of those physicists who so often remark that uncertainty appears only in the quantum realm. *The UFO Verdict* by Robert Sheaffer contains the premature certainty implied by the title. Mr. Sheaffer knows what UFOs "really are" — they "really are" hoaxes and hallucinations. Similarly, *Flying Saucers Are Real* by Major Donald Keyhoe contains the dogmatism implied by the title. Major Keyhoe also knows what UFOs "really are" — they "really are" interplanetary spaceships. The perception psychologist would note that UFOs come and go so fast, usually, that most of them never graduate from the *maybe* class. But we will expand on that in a more appropriate place.

Aristotelian dogmatic habit also reinforces and gets reinforced by ancient mammalian territorial imperatives. Wild primates, like other vertebrates, claim physical territories; domesticated primates (humans) claim "mental" territories — Ideologies and Religions. Thus, one seldom hears a quantum *maybe* in discussions of Roosevelt's economic policies versus Ronald Reagan's economic policies. Like Sheaffer and Keyhoe, most UFO believers and UFO debunkers disagree about everything else, but share a common aversion to the word *maybe*. And one virtually never hears *"Maybe* Jesus was the son of God" or *"Maybe* Islam is a false religion."

People ignore the quantum *maybe* because they have largely never heard of quantum logic or Transactional psychology, but they also ignore it because *traditional politics and religion have conditioned people for millenniums — and still train them today — to act with intolerance and premature certainty.*

In general, people judge it "manly" to pronounce dogmatic verdicts and fight for them, and to admit quantum uncertainty (von Neumann's *maybe)* seems "unmanly." Feminism often challenges this *machismo,* but, just as often, certain Feminists appear to think they will appear stronger if they speak and behave as dogmatically and unscientifically as the stupidest, most *macho* males.

This tendency to premature verdicts receives consider-able reinforcement, also, from the software our brains habitually use — our language and our typical language structure.

According to *News of the Weird* (by Shepherd, Cohut and Sweet, New American Library, 1989) — a book of nearly incredible but seemingly true stories from respectable newspapers — in 1987 a man in Rochester, New York, shot a woman he mistook for his wife. "I'm sorry," he told police. "I meant to shoot my wife, but I forgot my glasses." His universe, like those of Sheaffer and Keyhoe, seems built on quick verdicts and no *maybes*.

Same book — a man in Westchester shot his wife while hunting. He told police he had mistaken her for a wood-chuck.

Two more hunting accidents, in the same book — a man shot a friend, who he mistook for a squirrel. Another man shot a teen-age girl he mistook for a groundhog.

Another man in Virginia Beach killed his mother-in-law with a hatchet and claimed he mistook her for a large raccoon.

These stories have *maybes* in their *maybes*. I mean, after all, maybe some of these people made up these alibis out of desperation.

Then again, the world contains, not only UFOs (Uniden-tified Flying Objects) but also UNFOs (Unidentified Non-Flying Objects): and those without von Neumann's three-valued logic and its *maybes* will often be too quick about "understanding" and "identifying" them.

If you live in a busy part of a city, look out the window. Observe how many UNFOs go past so quickly that they never graduate from the *maybe* state to the "identified" state.

Exercizes

Classify the following propositions as true, false or maybe.

A. In 1933, Franklin Roosevelt became President of the United States.
B. In 1932, Franklin Roosevelt became President of the United States.
C. On January 18, 1932, Cary Grant had his 28th birthday.
D. The river Necker flows through the city of Frankfurt.
E. The river Necker flows through the city of Heidelberg.
F. Humanity evolved from Old World apes.
G. Force always equals mass multiplied by acceleration.
H. Francis Bacon wrote the plays attributed to Shakespeare.
I. Sex education leads to an increase in sex crimes.
J. In the years in which sex education increased in the U.S., reported sex crimes also increased.
K. The census of 1890 showed 4,000,000 inhabitants of New York City.
L. An ordinary pack of cigarettes contains 20 cigarettes.
M. Ronald Reagan knew about the Iran-Contra guns-and-cocaine deals of North, Secord and Hull.
N. Ronald Reagan did not know about the Iran-Contra crimes until he heard the news on TV.
O. All the differences between men and women result from cultural training.
P. Sombunall[1] of the differences between men and women result from cultural training.
Q. All differences between men and women result from genetic factors (testosterone, estrogen etc.)
R. Sombunall of the differences between men and women result from genetic factors (e.g., testosterone, estrogen etc.)
Q. The lost continent of Atlantis exists under the sea near Bermuda.
R. The lost continent of Atlantis never existed.
S. Hitler only had one testicle.

[1] Have you found this word useful in the week since doing the exercizes at the end of the last chapter?

NINE

How George Carlin Made Legal History

Everybody understands that you cannot drink the word "water", and yet virtually nobody seems entirely free of semantic delusions entirely comparable to trying to drink the *ink-stains* that form the word "water" on this page or the *sound waves* produced when I say "water" aloud. If you say, "The word is not the thing," everybody agrees placidly; if you watch people, you see that they continue to behave as if something called Sacred "really is" Sacred and something called Junk "really is" Junk.

This type of neurolinguistic "hallucination" appears so common among humans that it usually remains invisible to us, as some claim water appears invisible to fish, and we will continue to illustrate it copiously as we proceed. On analysis, this "word hypnosis" seems the most peculiar fact about the human race. Count Alfred Korzybski said we "confuse the map with the territory." Alan Watts said we can't tell the menu from the meal. However one phrases it, humans seem strangely prone to confusing their mental file cabinets — neurolinguistic grids — with the non-verbal world of sensory-sensual space-time.

As Lao-Tse said in the *Tao Te Ching*, 2500 years ago,

The road you can talk about is not the road you can walk on.

(Or:

The way that can be spoken is not the way that can be trodden.)

We all "know" this (or think that we do) and yet we all perpetually forget it.

For instance, here in the United States — an allegedly secular Democracy with an "iron wall" of separation between Church and State written into its Constitution — the Federal Communications Commission has a list of Seven Forbidden Words which nobody may speak on the radio or television. Any attempt to find out why these words remain *Tabu* leads into an epistemological fog, a morass of medieval metaphysics, in which concepts melt like Salvador Dali's clocks and ideas become as slippery as a boat deck in bad weather.

One cannot dismiss this mystery as trivial. When comedian George Carlin made a record ("Occupation: Foole") discussing, among other things, "The seven words you can never say on television," WBAI radio (New York) played the record, and received a fine so heavy that, although the incident occurred in 1973, WBAI, a small listener-sponsored station, recently announced (1990) that they have not yet paid all their legal costs in fighting the case, which went all the way to the Supreme Court. The Eight Wise Men (and One Wise Woman) thereon upheld the Federal Communications Commission.

The highest court in the land has actually ruled on what comedians may and may not joke about. George Carlin has become something more than a comedian. He now has the status of a Legal Precedent. You will pay a heavy fine, in the U.S. today, if you speak any of the Seven Forbidden Words on radio or television — *shit, piss, fuck, cunt, cocksucker, motherfucker* and *tits.*

The words have been forbidden, "our" Government says, because they "are" "indecent". Why "are" they "indecent"? Because a certain percentage of people who might turn on the radio or TV experience them as "indecent".

Why do sombunall people experience these words as "indecent"? Because the words "are" "dirty" or "vulgar".

Why "are" these words "dirty" and "vulgar" when other words, denoting the same objects or events, "are" not "dirty" or "vulgar"? Why, specifically, can a radio station be fined if a psychologist on a talk show says "He was so angry he wouldn't fuck her anymore" but not fined at all if the psychologist says "He was so angry he stopped having sexual intercourse with her"?

As Mr. Carlin pointed out in the comedy routine which led the Supreme Court to perform their even more remarkable comedy routine, fucking seems one of the most common topics on television, even though nobody uses the word. To paraphrase Mr. Carlin, many guests on the Merv Griffin and Donahue shows have written books on how to fuck or who to fuck or how to fuck better, and nobody objects as long as they say "sexual intercourse" instead of "fucking." And, of course, as Carlin goes on, the main topics on soap operas, day after day, consist of who has fucked whom, will she fuck him, will he fuck somebody else, have they fucked yet, who's getting fucked now, etc.

Some say "fuck" "is" "dirty" and "sexual intercourse" isn't because "fuck" comes from the Anglo-Saxon and "sexual intercourse" comes from the Latin. But then we must ask: how did Anglo-Saxon get to be "dirty" and why does Latin remain "clean"?

Well, others tell us, "fuck" represents lower-class speech and "sexual intercourse" represents middle-and-upper class speech. This does not happen to accord with brute fact, statistically: I have heard the word "fuck" in the daily (non-radio) conversation of professors, politicians, business persons, poets, movie stars, doctors, lawyers, police persons and most of the population of sombunall classes and castes, except a few religious conservatives.

And, even if "fuck" did occur exclusively in lower-class speech, we do not know, and can hardly explain, why it has been subject to a huge and bodacious fine when such other lower-class locutions as "ain't", "fridge" (for refrigerator), "gonna" and "whyncha" (why don't you) have not fallen under similar sanction. Nor have we yet seen a ban on the distinctly lower class "Jeet?" "Naw — Jew?" (Did you eat? No, did you?)

The fact that some enclaves of religious conservatives do not use the word "fuck" (or are embarrassed if they get caught using it) seems to provide the only clue to this mystery. The Federal Communications Commission, it seems, bases its policy upon persons who believe, or for political reasons wish to seem to believe, that the rather paranoid "God" of the conservative religions has His own list of Seven Forbidden Words and will become quite irate if the official Tabu list of our government does not match His list. Since that particular Deity has a reputation for blowing a few cities to hell whenever he feels annoyed, the F.C.C. may, in the back of their heads, think they will prevent further earthquakes by maintaining the Tabu on the Seven Unspeakable Words.

The Wall of Separation between Church and State, like many other pious pronouncements in our Constitution, does not correspond with the way our government actually functions. In short, the Seven Forbidden Words remain forbidden because pronouncing them aloud might agitate some Stone Age deity or other, and we still live in the same web of Tabu that controls other primitive peoples on this boondocks planet.

Some light seems about to dawn in the semantic murk...but let us press further and ask why the conservative's Stone Age "God" objects to "fuck" and not to "sexual intercourse" or such synonyms as "coitus", "copulation", "sexual congress", "sexual union", "love-making", etc.? Should we believe this "God" has a violent prejudice against words which, in reputation if not in reality, seem to reflect lower-class culture? Does this "God" dislike poor people as much as Ronald Reagan did?

Perhaps the reader will appreciate the immensity of this mystery more fully if I ask a related question:

If the word "fuck" "is" obscene or "dirty", why isn't the word "duck" 75% "dirty"?

Or, similarly:

If the word "cunt" "is" unacceptable to the conservative's "God", why does the word "punt" not receive a 75%

unacceptability rating? Why do we not see it spelled "p--" in the daily press?

To quote the admirable George Carlin one more time, "Such logic! Such law!"

Exercizes

1. Try to explain the difference between a *Playboy* center-fold and a nude by Renoir. Discuss among the whole group and see if you can arrive at a conclusion that makes sense when stated in operational-existential language.

2. Perform the same delicate semantic analysis upon a soft-core porn movie and a hard-core porn movie. Remember: try to keep your sentences operational, and avoid Aristotelian essences or spooks.

3. When U.S. troops entered Cambodia, the Nixon administration claimed this "was not" an invasion, because it "was only" an incursion. See if anybody can restate this difference in operational language.

4. The C.I.A. refers to certain acts as "termination with maximum prejudice." The press describes these acts as "assassinations." Try to explain to each other the difference.

Also, imagine yourselves as the victims. Do you care deeply whether your death gets called "termination with maximum prejudice" or "assassination"?

5. In the 1950s, the film *The Moon Is Blue,* became a center of controversy and actually got banned in some cities because it contained the word "virgin." How does this seem in retrospect? Discuss. (If anybody finds Mr. Carlin's paraphrased jokes offensive let them explain why the above film no longer seems offensive.)

TEN

Fussy Mutts & a City With Two Names

We have pointed out that nobody would try to drink the ink-stains that form the word "water" on this page, yet most people have semantic delusions and hallucinations entirely similar to that. The reader has perhaps begun to appreciate that this hardly qualifies as a hyperbole or exaggeration.

To paraphrase Professor S. I. Hayakawa, when you go into a restaurant you expect the menu to say "Choice Cut of Top Sirloin Steak" and would be taken aback if it said "A hunk of meat chopped off a dead castrated bull." Yet the two verbal formulae refer to same non-verbal event in the space-time continuum, as vegetarians would quickly tell us.

Words *do not equal in space-time* the things or events they denote, yet people react to a choice between words as if making a choice between "real" things or events in the existential world.

This "hypnosis by words" or "living in a cocoon of words" can even lead to murder. Literally. I recall three typical examples:

1. Several years ago in San Francisco, a man ordered an extra portion of steak in a restaurant, saying he wanted to take some home for his dog. The waiter commented that he personally fed his own dog on Red Heart dog food. The customer replied that his dog would not eat dog food and demanded steak. The waiter said, "You've got one fussy mutt there, mister." The man, badly hurt by these insensitive words, went home and brooded.

His beloved dog, a prince of canines, had been called "a fussy mutt." He brooded more. Imagine how you would feel if your mother were called "a drunken old whore." To this man, having his precious hound called "a fussy mutt" apparently seemed equally insufferable. He went back to the restaurant and shot the waiter dead.

2. Salman Rushdie recently composed the kind of orchestration of words and meanings we generally call a "novel", as distinguished from a "poem", an "insurance policy" or a "political speech." The late rev. Ayatollah Khomeni found this artistic arrangement of words as unbearable as the Federal Communications Commission finds the Seven Unspeakable Words on their Tabu List. As you have no doubt read, the rev. Ayatollah offered a reward of $5,000,000 to whoever would go to England and shoot Mr. Rushdie in the head with bullets.

(Mr. Rushdie had not referred to Mohammed as "a fussy mutt," but what he did write, even if intended as art, impacted upon the rev. Ayatollah as just as hurtful as "fussy mutt" to the man in San Francisco.)

3. When the English conquered Ireland, they changed the name of the old town of Derry to Londonderry. This has proven unacceptable to many Irish patriots and utterly insufferable to the Irish Republican Army. The Protestants in the town, on the other hand, prefer "Londonderry" to "Derry." As of 1990, if you say "Derry" in one part of that town you may well get shot by the Ulster Freedom Fighters, and if you say "Londonderry" on the other side of town you may get shot by the I.R.A.

The U.F.F. believe their fight "is" for the rights of the Protestant majority in Northern Ireland; the I.R.A. believe their fight "is" for the rights of the Catholic minority. Whose civil rights seem realistically infringed when one says "Derry" instead of "Londonderry" or "Londonderry" instead of "Derry"?

If I write that what actually exists in sensory-sensual space-time "is not" "Derry" or "Londonderry" but a collection of people, houses, parks, bridges, pubs, streets etc., I may seem to escape Ideology and move closer to existential "reality" or common human experience. Right? Wrong. On closer examination, this proves not quite the case. "A collection of people, houses, streets etc." consists of *words* and what you will find in that place in space-time remains *something not words* but non-verbal "things" and events.

"Non-verbal things and events," however, remains still words…in English…and we seem to have landed in another kind of Strange Loop.

Perhaps Zen Buddhism can enlighten us. After all, Zen has promised Enlightenment for several hundred years now.

A Zen *koan* of long standing goes as follows: The *roshi* (Zen teacher) holds up a staff and says, "If you call this a staff, you affirm. If you say it is not a staff, you deny. Beyond affirmation or denial, what is it?"

Exercize

I suggest that readers reflect on what has been said so far, about the Seven Forbidden Words and the "fussy mutt" and the killings in Northern Ireland. Reflect on "the map is not the territory" and "the menu is not the meal." Close the book, close your eyes, sit quietly, and think about this Zen riddle. Wait a minute and see if a light slowly dawns on you.

ELEVEN

What Equals the Universe?

Hello. Welcome back. Whether you have solved the staff riddle or not, I now invite you to again consider the question: What equals the universe?

According to pious Roman Catholics, the philosophy of Thomas Aquinas equals the universe. In other words, everything in the universe exists also in the philosophy of Aquinas and everything in Aquinas exists also in the universe.

On the other hand, according to pious Russian Communists, the ideology of Dialectical Materialism as developed by Marx, Engels and Lenin equals the universe. Everything in the universe exists in Dialectical Materialism and everything in Dialectical Materialism exists also in the universe.

Disciples of Ayn Rand feel the same way about Objectivism.

However, aside from Catholics, Marxists, Objectivists and a few other dingbat groups like the Committee for Scientific Investigation of Claims of the Paranormal or the Hard Shell Baptists, most of us, in this technological age, have at least a dim awareness that no coordination of words, however skillfully orchestrated, quite equals the whole universe. Whether we have ever stated it in the words I used in Part One or not, we have come to realize that nothing *equals* the universe except the universe itself.

Any philosophy, any theology, any coordination of words, any mathematical model, any scientific "system" must always remain something less than the whole universe. Such maps or models may describe large parts of the universe, but none of them can contain the whole universe. The most advanced mathematical physics, for instance, cannot predict what I will write in the next five minutes. (Neither can I, as Bergson pointed out.)

Some maps seem to contain fairly large areas of fiction, also — a possibility we always remember when considering other people's ideas but quickly forget when considering our own

Our maps and models of the universe — our reality-tunnels — always contain something less than a life-size working model of the universe, which we have not built yet and probably never can build. Such a life-size model, after all, would have to include you and me and every other sentient being throughout space-time and the sensations and/or thoughts of all such beings. Since nobody knows enough to build such a working model, nobody understands the universe in full.

But you realized that already if you figured out the answer to the Zen staff riddle. Right? However, if you still feel discombobulated...

Try it this way:

Not only does the word "water", an ink-stain on the page, not equal the experience of water, but our ideas about water will never contain all possible human experiences of water. The chemical formula for water, H_2O, tells us sombunall of what a chemist normally needs to know about water — namely that it has two hydrogen atoms for every oxygen atom — but it does NOT tell us, or attempt to tell us, the difference between drinking a glass of water on a hot day and being on a boat hit by several tons of rapidly moving water during a tropical storm. Nor does it tell us the difference between the role water plays in the life of a goldfish, who cannot survive more than about two minutes without water, and a human, who can survive nearly a week without water, but not much longer than that.

Brief Exercize

Meditate on the difference between the two sentences following, and note how coding (typographical convention) helps us distinguish the two meanings:

1. Water is not a word.
2. "Water" is a word.

Got it? No, you probably haven't. Not yet. You only think you've got it...

Exercizes

1. Let everybody in the class pinch their upper arms.

2. Let everybody then speak the word "pinch" out loud.

3. Let everybody write the word "pinch" on a piece of paper.

4. Let everybody again pinch their upper arms.

5. Discuss the differences among Exercizes 1 to 4.

TWELVE

The Creation of Reality-Tunnels

Our models of the universe — our glosses or gambles — have at least the following limitations and constraints upon them.

1. Genetics. Our DNA happens to have evolved out of standard primate DNA and still has a 98 per cent similarity to chimpanzee DNA (and 85% similarity to the DNA of the South American spider monkey).

We thus have basically the same gross anatomy as other primates, the same nervous systems, basically the same sense organs, etc. (Our more highly developed cortexes allow us to perform certain "higher" or more complex mental functions than other primates, but our perceptions remain largely within the primate norm.)

The DNA and the sensory/neural apparatus produced by the DNA creates what ethologists call the *umwelt* (world-field) perceived by an animal.

Dogs and cats see and sense a different *umwelt* or reality-tunnel than primates. (Hence, Heinlein's Law: "No way has ever been found to out-stubborn a cat.") Canine, feline and primate reality-tunnels, however, remain similar enough that friendship and communication among canines, felines and primates occurs easily.

Snakes live in a much different *umwelt*. They see heat waves, for instance, and seemingly do not see "objects." The world seen by a snake looks fundamentally like a spiritualist seance — fields of "life-energy" floating about in murk. That explains why a snake will strike at a hot air balloon that invades its territory. To the snake, the heat in the balloon and the heat in a hunter's leg have the same meaning — an intruder has arrived. The snake defends its territory by striking in both cases.

Because the snake's *umwelt* or reality-tunnel differs so fundamentally from mammalian reality tunnels, friendships between humans and snakes occur much less frequently than friendships between humans and mammals.

The belief that the human *umwelt* reveals "reality" or "deep reality" seems, in this perspective, as naive as the notion that a yardstick shows more "reality" than a voltmeter, or that "my religion 'is' better than your religion." *Neurogenetic chauvinism has no more scientific justification than national or sexual chauvinisms.*

The most ingenious recent attempt to revive classical Aristotelianism occurs in Anthony Stevens's book, *Archetypes,* which argues that evolution must have produced sense organs that reveal "truth" or "reality" or something of the sort, or we would have become extinct by now. This argument overlooks several facts; *viz...*

More species have become extinct than currently survive.

Most extinct species died of their own limitations before humans arrived, and their demise did not result from "human rapacity."

Many tribes of humans have become extinct.

Whole civilizations have destroyed themselves, some clearly by following crazy deductions from inadequate perceptions.

Considering evolution with these facts in mind, we see that most animals perceive as accurate a reality-tunnel *of their local habitat* as will statistically allow most members of that species to survive long enough to reproduce. No animal, including the domesticated primate, can smugly assume the world revealed/created by its senses and brain equals in all respects the real world or the "only real world." It equals sombunall, not all.

2. Imprints. It appears (as of 1990) that animals have brief periods of imprint vulnerability in which their nervous systems can suddenly create a personalized reality-tunnel unique to itself. These imprints permanently bond neurons into reflex networks which seemingly remain for life. The basic research on imprinting, for instance, for which Lorenz and Tinbergen shared a Nobel Prize in 1973, demonstrated that the statistically normal snow-goose imprints its mother, as distinct from any other goose, shortly after birth. This imprint creates a "bond" and the gosling attaches itself to the mother in every way possible.

These brief points of imprint vulnerability can literally imprint anything. Lorenz, for instance, recorded a case in which a gosling, in the temporary absence of the mother, imprinted a ping-pong ball. It followed the ping-pong ball about, nestled with it, and, on reaching adulthood, attempted to mount the ball sexually. Another gosling imprinted Dr. Lorenz himself, with equally bizarre results.

In any litter of puppies, you can easily observe how rapidly Top Dog and Bottom Dog roles get imprinted. The Top Dog eats more, grows larger and continues as Top Dog for life; the Bottom Dog remains submissive and "timid."

A quick examination of any human community lends plausibility to Dr. Timothy Leary's hypothesis that most humans have imprinted Top Dog or Bottom Dog roles just as mechanically as canines do. (In sombunall mammals, of course, some individuals imprint roles midway between Top and Bottom, and thus hierarchy appears...)

How and when we imprint language seems to determine lifelong programs of "cleverness" (verbal facility) or "dumbness" (verbal clumsiness). This reflects in our speech, and, since thinking consists largely of juggling words sub-vocally, in our ability to handle concepts, solve problems, etc.

How and when our pubertal sexuality gets imprinted, similarly, seems to determine lifelong programs of heterosexuality or homosexuality, brash promiscuity or monogamy, etc. In both common sexual imprints like these and in more eccentric imprints (celibacy, foot fetishism, sadomasochism etc.) the bonded brain circuitry seem quite as mechanical as the imprint which bonded the gosling to the ping-pong ball. (Anybody who doubts this can try responding sexually to a stimulus that never turned them on before, or totally ignoring a stimulus that normally does turn them on.)

Thus, nobody enters a room with their genetic primate neurology as the only constraint on what they will perceive. Depending on imprints, one may "see" from a clever heterosexual Top Dog position, from a clever homosexual Top Dog perspective, from a dumb homosexual Top Dog position, from a dumb heterosexual Bottom Dog perspective, from a smart celibate Bottom Dog vantage etc. etc. The permutations and possibilities seem quite large, although finite.

Genetics and hard-wired imprints do not make up the whole of the software which programs our selves and our perceived universes. There remain:

3. Conditioning. Unlike imprinting which requires only one experience and sets permanently into the neurons, conditioning requires many repetitions of the same experience and does not set permanently. Behaviorists, also, know how to reverse conditioning by counter-conditioning, but only Dr. Timothy Leary has claimed to know how to reverse imprinting. (Curiously, laws currently forbid other scientists to repeat and test Dr. Leary's experiments, and threaten them with prison if they do get caught repeating this research. The idea that the Inquisition died 170 years ago seems, like the Separation of Church and State, just another myth, unconnected with how "our" government actually functions.)

4. Learning. Like conditioning, this requires repetition and it also requires *motivation.* For these reasons, it seems to play less of a role in human perception and belief than genetics and imprinting do and even less than conditioning does.

It seems, as of 1990, that all snakes perceive virtually the same reality-tunnel, with only minor imprinted differences. Mammals show more conditioned and learned differences in their reality-tunnels. (Most "clever dog" stories that get into newspapers illustrate that some particular dog has imprinted a model of the world unlike that of any other dog we have known.)

Humans, due to our complicated cortex and frontal lobes, which permit more conditioning and learning, and also due to our prolonged infancy (which probably permits more imprints, and more eccentric imprints) seem to differ more from each other than any other mammals. Thus, an Irish dog, an Afghani dog, a Russian dog etc., generally understand each other fairly well. The canine reality-tunnel has more commonalities than differences, as we have said. However, an Irishman wrapped in a Catholic reality-tunnel and a Top Dog personality may have great difficulty understanding an Afghani living in a Moslem reality-tunnel with a Bottom Dog personality, and both may find it impossible to communicate at all with a Russian homosexual communist Top Dog.

This variability of humans can function as the greatest evolutionary strength of the human race, since it may *allow us to learn from persons imprinted and/or trained to see and hear and smell and think those things we have learned not to see or hear of smell or think.*

Due to our habit of premature certainty, however, this variability seldom serves that beneficial evolutionary function. More often, when meeting somebody with a different gloss or *umwelt,* we merely label that person "mad" or "bad" — crazy or evil — or both.

This may explain most of the hostility on this planet, and most of the wars.

Apologists for certain authoritarian/dogmatic groups (the Vatican, the U.S. State Department, the Politburo, CSICOP) spend most of their time constructing "proofs" that anybody who does not share their reality-tunnel has serious mental or moral defects or "is" a damned liar.

Again: I call this book Quantum Psychology rather than Quantum Philosophy because *understanding and internalizing (learning to use) these principles can decrease dogma, intolerance, compulsive behavior, hostility, etc. and also may increase openness, continuous learning, "growth" and empathy —*

sombunall of which represent goals sought in most forms of psychotherapy, and sombunall forms of mystic religion.

Exercizes

1. Let one member of the study group write to the Flat Earth Research Society, Box 2533, Lancaster, CA 93539. Let him or her present to the group some good arguments that the flatness model of Earth fits the facts better than the sphere model.

Let all members attempt to listen calmly, rationally, objectively.

Let all members observe that this attempt to listen without prejudice seems very much more difficult than you would expect in advance.

2. Let another member of the group similarly research and present a defense of Islam (especially its attitude toward women).

Again, attempt to listen without prejudice, and observe how hard this seems.

3. Let another member research the brilliant scientist Nikola Tesla, father of alternating current grids, and present to the group Tesla's reasons for rejecting Relativity.

4. Let another member research and present the case against Evolution.

Again: More profit will come from doing these exercizes than from merely reading about them.

THIRTEEN

E and E-Prime

In 1933, in *Science and Sanity*, Alfred Korzybski proposed that we should abolish the "is of identity" from the English language. (The "is of identity" takes the form *X is a Y*. E.g., "Joe is a Communist," "Mary is a dumb file-clerk," "The universe is a giant machine," etc.) In 1949, D. David Bourland Jr. proposed the abolition of all forms of the words "is" or " to be" and the Bourland proposal (English without "isness") he called E-Prime, or English-Prime.

A few scientists have taken to writing in E-Prime (notably Dr. Albert Ellis and Dr. E.W. Kellogg III). Bourland, in a recent (not-yet-published) paper tells of a few cases in which scientific reports, unsatisfactory to sombunall members of a research group, suddenly made sense and became acceptable when re-written in E-Prime. By and large, however, E-Prime has not yet caught on either in learned circles or in popular speech.

(Oddly, most physicists write in E-Prime a large part of the time, due to the influence of Operationalism — the philosophy that tells us to define things by operations performed — but few have any awareness of E-Prime as a discipline and most of them lapse into "isness" statements all too frequently, thereby confusing themselves and their readers.)

Nonetheless, E-Prime seems to solve many problems that otherwise appear intractable, and it also serves as an antibiotic against what Korzybski called "demonological thinking". Most of this book employs E-Prime so the reader could begin to get acquainted with this new way of mapping the world; in a few instances I allowed normal English, and its "isness" to intrude again (how many of you noticed that?), while discussing some of the weird and superstitious thinking that exists throughout our society *and always occurs when "is" creeps into our concepts.* (As a clue or warning, I placed each "is" in dubious quotation marks, to highlights its central role in the confusions there discussed.)

As everybody with a home computer knows, *the software can change the functioning of the hardware in radical and sometimes startling ways.* The first law of computers — so ancient that some claim it dates back to dark, Cthulhoid aeons when giant saurians and Richard Nixon still dominated the earth — tells us succinctly, "Garbage In, Garbage Out" (or GIGO, for short).

The wrong software *guarantees* wrong answers, or total gibberish. Conversely, the correct software, if you find it, will often "miraculously" solve problems that had hitherto appeared intractable.

Since the brain does not receive raw data, but edits data as it receives it, we need to understand the software the brain uses. The case for using E-Prime rests on the simple proposition that "isness" sets the brain into a medieval Aristotelian framework and makes it impossible to understand modern problems and opportunities. A classic case of GIGO, in short. Removing "isness" and writing/thinking only and always in operational/existential language, sets us, conversely, in a modern universe where we can successfully deal with modern issues.

To begin to get the hang of E-Prime, consider the following two columns, the first written in Standard English and the second in English Prime.

Standard English	English Prime
1. The photon is a wave.	1. The photon behaves as a wave when constrained by certain instruments.
2. The photon is a particle.	2. The photon appears as a particle when constrained by other instruments.
3. John is unhappy and grouchy.	3. John appears unhappy and grouchy in the office.
4. John is bright and cheerful.	4. John appears bright and cheerful on holiday at the beach.
5. The car involved in the hit-and-run accident was a blue Ford.	5. In memory, I think I recall the car involved in the hit-and-run accident as a blue Ford
6. That is a fascist idea.	6. That seems like a fascist idea to me.
7. Beethoven is better than Mozart.	7. In my present mixed state of musical education and ignorance Beethoven seems better than Mozart to me.
8. Lady Chatterley's Lover is a pornographic novel.	8. Lady Chatterley's Lover seems like a pornographic novel to me.
9. Grass is green.	9. Grass registers as green to most human eyes.
10. The first man stabbed the second man with a knife.	10. I think I saw the first man stab the second man with a knife.

In the first example a "metaphysical" or Aristotelian formulation in Standard English becomes an operational or existential formulation when rewritten in English Prime. This may appear of interest only to philosophers and scientists of an operationalist/phenomenologist bias, but consider what happens when we move to the second example.

Clearly, written in Standard English "The photon is a wave" and "The photon is a particle" contradict each other, just like the sentences "Robin is a boy" and "Robin is a girl." Nonetheless, all through the nineteenth century physicists found themselves debating about this and, by the early 1920s, it became obvious that the experimental evidence could not resolve the question, since the experimental evidence depended on the instruments or the instrumental set-up (design) of the total experiment. One type of experiment always showed light traveling in waves, and another type always showed light traveling as discrete particles.

This contradiction created considerable consternation. As noted earlier, some quantum theorists joked about "wavicles". Others proclaimed in despair that "the universe is not rational" (by which they meant to indicate that the universe does not follow Aristotelian logic). Still others looked hopefully for the definitive experiment (not yet attained in 1990) which would clearly prove whether photons "are" waves or particles.

If we look, again, at the translations into English-Prime, we see that no contradiction now exists at all, no "paradox", no "irrationality" in the universe. We also find that we have constrained ourselves to talk about what actually happened in space-time, whereas in Standard English we allowed ourselves to talk about something that has never been observed in space-time at all — the "isness" or "whatness" or Aristotelian "essence" of the photon. (Niels Bohr's Complementarity Principle and Copenhagen Interpretation, the technical resolutions of the wave/particle duality within physics, amount to telling physicists to adopt "the spirit of E-Prime" without quite articulating E-Prime itself.)

The weakness of Aristotelian "isness" or "whatness" statements lies in their assumption of indwelling "thingness" — the assumption that every "object" contains what the cynical German philosopher Max Stirner called "spooks". Thus in Moliere's famous joke, an ignorant doctor tries to impress some even more ignorant lay persons by "explaining" that opium makes us sleepy because it has a "sleep-producing property" in it. By contrast a scientific or operational statement would define precisely how the *structure* of the opium molecule chemically bonds to specific receptor *structures* in the brain, describing actual events in the space-time continuum.

In simpler words, the Aristotelian universe assumes an assembly of "things" with "essences" or "spooks" inside them, where the modern scientific (or existentialist) universe assumes a network of structural relationships. (Look at the first two samples of Standard English and English Prime again, to see this distinction more clearly.)

Moliere's physician does not seem nearly as comical as the theology promulgated by the Vatican. According to Thomist Aristotelianism (the official Vatican philosophy) "things" not only have indwelling "essences" or "spooks" but also have external "accidents" or appearances. This "explains" the Miracle of the Transubstantiation. In this astounding, marvelous, totally wonderful, even mind-boggling Miracle, a piece of bread changes into the body of a Jew who lived 2000 years ago.

Now, the "accidents" — which include everything you can observe about the bread, with your senses or with the most subtle scientific instruments — admittedly do not change. To your eyes or taste buds or electron microscopes the bread has undergone no change at all. It doesn't even weigh as much as a human body, but retains the weight of a small piece of bread. Nonetheless, to Catholics, after the Miracle (which any priest can perform) the bread "is" the body of the aforesaid dead Jew, one Yeshua ben Yusef, whom the goys in the Vatican call Jesus Christ. In other words, the "essence" of the bread "is" the dead Jew.

It appears obvious that, within this framework, the "essence" of the bread can "be" anything, or can "be" asserted to "be" anything. It could "be" the essence of the Easter Bunny, or it could "be" Jesus and the Easter Bunny both, or it could "be" the Five Original Marx Brothers, or it could "be" a million other spooks happily co-existing in the realm outside space-time where such metaphysical entities appear to reside.

Even more astounding, this Miracle can only happen if the priest has a Willy. Protestants, Jews, Zen Buddhists etc. have ordained many female clergy-persons in recent decades, but the Vatican remains firm in the principle that only a male — a human with a Willy — can transform the "essence" of bread into the "essence" of a dead body.

(Like the cannibalism underlying this Rite, this phallus-worship dates back to Stone Age ideas about "essences" that can be transferred from one organism to another. Ritual homosexuality, as distinguished from homosexuality-for-fun, played a prominent role in many of the pagan fertility cults that got incorporated into the Catholic metaphysics. See Frazer's *Golden Bough* and Wright's *Worship of the Generative Organs.* It requires a phallus to transmute bread into flesh because our early ancestors believed it requires a phallus to do any great work of Magick.)

In Standard English we may discuss all sorts of metaphysical and spooky matters, *often without noticing that we have entered the realms of theology and demonology,* whereas in English Prime we can only discuss actual experiences (or transactions) in the space-time continuum. English Prime may not automatically transfer us into a scientific universe, in all cases, but it at least transfers us into existential or experiential modes, and takes us out of medieval theology.

Now, those who enjoy theological and/or demonological speculations may continue to enjoy them, as far as I care. This book merely attempts to clarify the difference between theological speculations and actual experiences in space-time, so that we do not wander into theology without realizing where we have gotten ourselves. The Supreme Court, for instance, wandered into theology (or demonology) when it proclaimed that "fuck" "is" an indecent word. The most one can say about that in scientific E-Prime would read: "The word 'fuck' appears indecent in the evaluations of X per cent of the population," X found by normal polling methods.

Turning next to the enigmatic John who "is" unhappy and grouchy yet also "is" bright and cheerful, we find a surprising parallel to the wave/particle duality. Remaining in the reality-tunnel of Standard English, one might decide that John "really is" manic-depressive. Or one speaker might decide that the other speaker hasn't "really" observed John carefully, or "is" an "untrustworthy witness." Again, the innocent-looking "is" causes us to populate the world with spooks, and may provoke us to heated debate, or violent quarrel. (That town in Northern Ireland mentioned earlier — "is" it "really" Derry or Londonderry?)

Rewriting in English Prime we find "John appears unhappy and grouchy in the office" and "John appears bright and cheerful on holiday at the beach." We have left the realm of spooks and re-entered the existential or phenomenological world of actual experiences in space-time. And, lo and behold, another metaphysical contradiction has disappeared in the process.

To say "John is" *anything*, incidentally, always opens the door to spooks and metaphysical debate. The historical logic of Aristotelian philosophy as embedded in Standard English always carries an association of stasis with every "is", *unless the speaker or writer remembers to include a date*, and even then linguistic habit will cause many to "not notice" the date and assume "is" means a stasis (an Aristotelian timeless essence or spook).

For instance, "John is beardless" may deceive many people (but not trained police officers) if John becomes a wanted criminal and alters his appearance by growing a beard.

"John is a Protestant" or "John is a Catholic" may change any day, if John has developed a habit of philosophical speculation.

Even stranger, "John is a Jew" has at least five different meanings, some of which may change and some remain constant, and only one of which tells us anything about how John will behave in space-time.

Perhaps I'd better enlarge on that last point. "John is a Jew," according to Rabbinical law, means that John had a Jewish mother. This tells us nothing about John's politics or religion, and less than nothing about his taste in art, his sexual life, his favorite sports etc.

"John is a Jew" in Nazi Germany, or in anti-Semitic enclaves in the U.S. today, means that John had one known ancestor somewhere who could be classified as "Jewish" by one of these five contradictory definitions. Again, this tells us nothing about how John will behave.

"John is a Jew" in some circles means that John practices the Jewish religion. At last we have learned something about John. He will certainly attend Temple regularly...or fairly regularly. (But we still don't know how strictly he will follow the kosher laws...)

"John is a Jew" in some other circles means that, while John rejects the Jewish religion, he identifies with "the Jewish community" and (if he has become famous) might speak "as a Jew" at a political rally. (We still don't know, e.g., whether he will support or criticize current Israeli policies.)

"John is a Jew" can also mean that John lives in a society where, for any one of the above reasons, people regard him as a Jew, and he perforce has to recognize this "Jewishness" as something — even if only a spook — that people usually "see" when they think they see *him*.

Literary critics, usually considered careful and analytical readers, or more careful and analytical that most, referred to Leopold Bloom, the hero of James Joyce's *Ulysses,* as a "Jew" for over 40 years. Only in the last decade or so have Joyce scholars begun arguing about whether Bloom "is" a Jew or not. (Bloom qualifies as Jewish in only two of the five meanings above and appears not-Jewish in three. Does that make him "40% Jewish" or 60% "not-Jewish"? Or both?) The emerging consensus of Joycean studies now appears to recognize that Joyce gave Bloom a very tangled genetic/cultural background just to create this ambiguity and thereby satirize anti-Semitism more sharply.

I may seem eccentric to suggest that, without formulating E-Prime explicitly, Joyce, like his great contemporary, Bohr, wished us to see beyond the fallacy contained in "isness" statements. Just like Schroedinger's cat ("dead" in some *eigenstates*, "alive" in others) Bloom does not make sense as a man in an environment until we recognize that both his "Jewishness" and his "non-Jewishness" play roles in his life, at different times, within different environments.

Incidentally, within the structure of Standard English, "Marilyn Monroe was a Jew" qualifies as correct, *although dated*, even though she had no known Jewish ancestors, no Jewish mother, did not show much "community feeling" with other Jews, and hardly ever got called a Jew in print. Nonetheless, while married to Arthur Miller, Marilyn practiced the Jewish religion and therefore in Standard English "was" more of a Jew than some of my atheist friends of Jewish ancestry. But returning to John...

"John is a plumber" also contains a fallacy. John may have quit plumbing since you last saw him and may work as a hair dresser now. Stranger things have happened. In E-Prime one would write "John had a job as plumber the last I knew."

Trivial? Overly pedantic? According to a recent article[1] Professor Harry Weinberg — curiously, an old acquaintance of mine — once tried to emphasize these points to a class by trying to make them see the fallacy in the statement "John F. Kennedy is President of the United States." Dr. Weinberg pointed out that the inference, *Nothing has changed since we came into this classroom*, had not been checked by anybody who insisted the statement about Kennedy contained certainty. Weinberg, like his students, got the lesson driven home with more drama that anybody expected, because this class occurred on November 22, 1963, and everybody soon learned that during that class time John F. Kennedy had died of an assassin's bullet and Lyndon B. Johnson had taken the oath as President of the United States.

---

1 "Statement of Fact or Statement of Inference" by Ruth Gonchar Brennan, *Temple Review*, Temple University, Winter 1988-89.

That makes the idea kind of hard to forget, doesn't it?

Looking at sample five — "The car...was a blue Ford" we might again encounter Bertrand Russell's two-head paradox. It seems a blue Ford exists "in" the head of the witness, but whether the blue Ford also existed "outside" that head remains unsure. Even outside tricky psychology labs, ordinary perception has become problematical due to the whole sad history of eye-witness testimony frequently breaking down in court. Or does the "external universe" (including the blue Ford) exist in some super-Head somewhere? It seems that the translation into E-Prime — "I recall the car...as a blue Ford" better accords with the experiential level of our existence in space-time than the two heads and other paradoxes we might encounter in Standard English.

James Thurber tells us that he once saw an admiral, wearing a 19th Century naval uniform and old-fashioned side-whiskers, peddling a unicycle down the middle of Fifth Avenue in New York. Fortunately, Thurber had broken his glasses and had not yet received replacements from the optometrist, so he did not worry seriously about his sanity. In the Castro section of San Francisco, a well-known homosexual area, I once saw a sign that said "HALF GAY CLEANERS" — but when I looked again, it said "HALF DAY CLEANERS".

Even Aristotle, despite the abuse he has suffered in these pages, had enough common sense to point out, once, that "I see" always contains fallacy; we should say "I have seen." Time always elapses between the impact of energy on the eye and the *creation* of an image (and associated name and ideas) in the brain, which explains why three eyewitnesses to a hit-and-run such as we postulate here may report, not just the blue Ford of the first speaker, but a blue VW or maybe even a green Toyota.

I once astonished a friend by remarking, *apropos* of UFOs, that I see two or three of them a week. As a student of Transactional Psychology, this does not surprise or alarm me. I also see UNFOs, as noted earlier — and I do not rush to identify them as raccoons or groundhogs, like some people we met earlier. Most people see UNFOs, without thinking about the implications of this, especially when driving rapidly, but sometimes even when walking. We only find UFOs impressive because some people claim they "are" alien spaceships. My UFOs remain Unidentified, since they did not hang around long enough for me to form even a guess about them, but I have found no grounds for classifying them as space-ships. Anybody who does not see UFOs frequently, I think, has not mastered perception psychology or current neuroscience. The sky contains numerous things that go by too quickly for anybody to *identify* them.

My own wife has appeared as an UNFO to me on occasion — usually around two or three in the morning when I get out of bed to go to the john and then encounter a Mysterious and Unknown figure emerging from the dark at the other end of the hall. In those cases, fortunately, identification did not take long, and I never reached for a blunt instrument to defend myself. Whatever my critics may suspect, I never mistook her for a squirrel.

If you think about it from the perspective of E-Prime, the world consists mostly of UFOs and UNFOs. Very few "things" (space-time events) in the air or on the ground give us the opportunity to "identify" them with certainty.

In example six — "That is a fascist idea" versus "That seems like a fascist idea to me" — Standard English implies an indwelling essence of the medieval sort, does not describe an operation in space-time, and mentions no instrument used in measuring the alleged "fascism" in the idea. The English Prime translation does not assume essences or spooks, describes the operation as occurring in the brain of the speaker and, implicitly, identifies said brain as the instrument making the evaluation. Not accidentally, Standard English also assumes a sort of "glass wall" between observer and observed, while English Prime draws us back into the modern quantum world where observer and observed form a seamless unity,

In examples 7 and 8, Standard English again assumes indwelling spooks and continues to separate observer from observed; English Prime assumes no spooks and reminds us of **QUIP** (the **QU**antum Inseparability Principle, so named by Dr. Nick Herbert), namely, the impossibility of existentially separating observer and observed.

Meditating on example 9 will give you the answer to a famous Zen *koan*, "Who is the Master who makes the grass green?" It might also save you from the frequent quarrels (mostly occurring between husbands and wives) about whether the new curtains "are really" green or blue.

Example 10 introduces new subtleties. No explicit "is" appears in the Standard English, so even those trained in E-Prime may see no problem here. However, if the observation refers to a famous (and treacherous) experiment well-known to psychologists, the Standard English version contains a hilarious fallacy.

I refer to the experiment in which two men rush into a psychology class, struggle and shout, and then one makes a stabbing motion and the other falls. The majority of students, whenever that has been tried, report a knife in the hand of the man who made the stabbing (knife-wielding) motion. In fact, the man used no knife. He used a banana.

Look back at the re-translation into E-Prime. It seems likely that persons trained in E-Prime will grow more cautious about their perceptions and not "rush to judgment" in the manner of most of us throughout history. They might even see the banana, instead of hallucinating a knife?

Exercizes

1. Have the group experiment with rewriting the following Standard English sentences into English Prime. Observe carefully what disagreements or irritability may arise.

    A. "The fetus is a person."
    B. "The zygote is a person."

C. "Every sperm is sacred/Every sperm is great/If a sperm is wasted/God gets quite irate." (Monty Python)

D. "Pornography is murder." (Andrea Dworkin)

E. "John is homosexual."

F. "The table is four feet long."

G. "The human brain is a computer."

H. "When I took LSD, the whole universe was transformed."

I. "Beethoven was paranoid, Mozart was manic-depressive and Wagner was megalomaniac."

J. "Today is Tuesday."

K. "*Lady Chatterley's Lover* is a sexist novel."

L. "Mice, voles and rabbits are all rodents."

M. "The patient is resisting therapy."

N. "Sin and redemption are theological fictions. The sense of sin and the sense of redemption are actual human experiences." (Paraphrased from Ludwig Wittgenstein.)

2. Repeat the experiment of passing the rock around the group, with each person trying to sense and feel the rock without forming any words about the rock in their brains.

3. Let each member of the group contemplate the following sentences, then let each one pick out the sentence that she or he would find most embarrassing to say out loud:

A. My mother was a drunken whore.

B. I am a cock-sucking homosexual queer.

C. I am a dyke and I'm proud of it.

D. I have always been a coward.

E. I am afraid to be alone in the dark.

F. I would be very happy if my spouse dropped dead.

Let each member speak out loud the sentence that arouses the most emotional resistance.

Let other members observe the tone and "body language" of the person trying to say something he or she dreads to say. Observe especially smiles (how sincere do they look?) or embarrassed giggles.

Let the members discuss the results of this. Especially, let them discuss why, after studying a chapter about the differences between words and non-verbal existence, most of us still fear certain words or ideas. And let them note how everybody probably showed (by tone, body-language etc.) that they did not "mean" what they said, as compared to the performance of a good actor who could speak any of these sentences with total conviction.

Recall the famous "penis" scene in the film, *Born on the Fourth of July* (in which Tom Cruise as a paralyzed veteran tries to explain to his mother what lifelong impotence means to him). Compare his "sincerity" and conviction in shouting that his penis will not get hard ever again with the comparative lack of "sincerity" of the class, who have presumably not had dramatic training.

How do actors learn to get beyond the taboos that control most of us? Do any of you, of heterosexual preference, think you could portray a homosexual as well as Brando once did? Why not? Discuss this in the group.

PART THREE

The Observer-Created Universe

Organized skepticism is a two-edged sword. It allows us to question orthodoxy as well as unorthodoxy... The scientist who claims to be a true skeptic, a zetetic, is willing to investigate empirically the claims of the American Medical Association as well as those of the faith-healer; and more important, he should be willing to compare the empirical results of both before defending one and condemning the other.   — Marcello Truzzi, Ph.D.

The Zetetic Scholar, Nos. 12-13 (1987)

FOURTEEN

The Farmer & The Thief

An old Chinese parable tells of a farmer who noticed that his coin purse had disappeared. Searching everywhere, he could not find the purse, and he became convinced that it had been stolen. Thinking back over who had visited his house recently, the farmer decided he knew who had stolen the purse — the son of a neighbor. The boy had visited the house the very day before the purse disappeared, and nobody else had had an opportunity to commit the burglary.

The next time the farmer saw the boy, he noticed plenty of behavioral "clues" to support his suspicious. The boy had a definitely furtive and guilty manner about him and, in general, looked as sneaky as a barn-rat. Knowing he had no legal proof, the farmer could not decide what to do. But every time he saw the boy after that, the guilty behavior of the lad increased markedly, and the farmer grew angrier. Finally, he felt so much anger that he decided to go to the boy's father and make a formal accusation.

Just then the farmer's wife called him. "Look what I found behind the bed," she said — and handed him the missing coin purse.

Taoist philosophers have often cited this parable, and point out that we can explain the innocent boy's "guilty" behavior in two ways:

1. Possibly, the boy never behaved in a way that would have looked guilty, "furtive," sly etc. to anybody else except the suspicious farmer. The farmer saw all these things only because he expected to see them.

That makes sense, even though it may undermine all our dogmas if we think about it deeply enough. If you ever noticed that, in a political quarrel, the people who agree with you seem justifiably angry at the tactics of the other side, while the other side seems "too bloody emotional to think straight" — or, if, like me, you once saw a sign saying "HALF GAY CLEANERS" in a homosexual neighborhood — or if you ever saw a creature that appeared half-horse and half-deer, like two men whose experience I recounted in *The New Inquisition* — you can understand that much of this parable.

Remember, also, the gent who shot his wife, convinced she "was" a squirrel.

Perhaps training in E-Prime and von Neumann's yes/no/maybe logic might prevent such "projections" (as the Freudians say)?

Let us remember that Transactional psychology has proven that, contrary to common sense and the prejudices of centuries, our minds do not passively receive impressions from the "external world." Rather we actively *create* our impressions: out of an ocean of possible signals, our brains notice the signals that fit what we expect to see, and we organize these signals into a model, or reality-tunnel, that marvelously matches *our ideas about what "is really" out there.*

One might that say the boy's guilt always remained in the "maybe" state existentially, until the purse reappeared, but the boy's guilt gradually moved beyond existential "maybe" to subjective certainty as the farmer's perceptions reinforced his suspicions.

Aristotle only noted that "I see" actually means "I have seen." Modern neuroscience reveals that "I see" (or "I perceive") actually means "I have made a bet." In the time between the arrival of signals at our eye or other receptor organ and the emergence of an image or idea in our brains, we have done a great deal of creative "artistic" work. We generally do that work so fast that we do not notice ourselves doing it. Thus, we forget the *gamble* in every perception and feel startled (or even annoyed) whenever we come up against evidence that others do not "see" what we "see."

Constant reminding ourselves that we do not see with our eyes but with our synergetic eye-brain system working as a whole will produce constant astonishment as we notice, more and more often, how much of our perceptions emerge from our preconceptions.

Training ourselves to write in E-Prime will vastly accelerate our progress in "internalizing" (learning to use) this modern knowledge. *Speaking* and *thinking* in E-Prime take much longer to learn, and relapses into "isness" occur 20 years, 40 years, or longer, after we think we have learned this lesson.

2. But Taoists also point out that the boy in the story may indeed have developed *some "guilty" behaviors* — fear of looking the farmer directly in the eye, for instance — just because he had become aware that the farmer suspected him of something.

We have entered the area of what behavioral science calls "self-fulfilling prophecies." This sort of thing happens to all of us. Some man repeatedly seems cold or unfriendly; we become a bit reserved ourselves, or try to avoid him. We have started to behave as he expected us to behave.

You, a male, have to have a business meeting with a Feminist who expects all, not sombunall, men to act unfairly or brutally. You try to remain calm and judicious, but her attitude annoys you more and more. Eventually, her hostility keys off your hostility. A prophecy has fulfilled itself: you have "proven" to her that her view of men as dangerous creatures has just had itself confirmed one more time.

Or: you, an Afro-American, confront a cop who "knows" all, not sombunall, Afro-Americans "are" violent and dangerous. He uses excessive force. You get angry and fight back. You have just confirmed your suspicion about white cops — and he has confirmed his suspicion about Afro-Americans.

Dealing with paranoids brings this circular-causal process into particularly sharp focus. No matter what

gambit you try, each act reinforces the paranoid's conviction that you, too, serve the Conspiracy that persecutes
him. Unless you work as a psychotherapist and have him
as a patient, you will eventually give up entirely and stop
trying to persuade him you have not joined any plots
against him: you just avoid him as much as possible. The
result? Your avoidance becomes one more item in his long
list of "proofs" of your guilt.

Medical researchers know that every innovative therapy
produces its best results when new, and some therapies
only produce results when new. For instance, the once-
touted "cancer cure," Krebiozen produced several notable
cures in its early days, but nobody has reported any cures
with Krebiozen in around 30 years now. In cases of this
sort, the enthusiasm of the researchers somehow communicates itself to the patients, who then "give themselves
permission" to get well.

Such cures-by-suggestion seem "miraculous" — or at
least "mysterious" — to most of us. This again indicates
word-hypnosis, or confusing the map with the territory.
We have two words in English (and related languages), "mind"
and "body", and we tend to think the universe must also have a
similar bifurcation. When we think in a more modern scientific language — in terms of the-organism-as-a-whole or
psychosomatic synergy etc. — such cures do not seem
either miraculous or mysterious.

Patients fed on a high dose of Optimism statistically fare
somewhat better than those fed only on grim Pessimism.
This should appear no more astounding than the recorded
fact that children who see a lot of violent horror movies
have more sleep disorders than children who see only
comedies.

If you treat a young man as a thief, he will begin to act
uncomfortable around you, which looks to the naked eye
much like a young man acting guilty. If a doctor expects
the patient to get well, this has some effect on the patient;
if the doctor expects the patient to die, this also has an
effect. Christian Science practitioners and other "faith-
healers" could not remain in business if such self-fulfilling

prophecies did not work out significantly often, statistically, in a variety of illnesses, sometimes even very serious illnesses.

Similarly, every imperialistic or conquering nation has proclaimed that the subject people "are" shiftless or lazy or dirty or ignorant or unreasonable or irrational or generally "inferior." Most subjugated peoples very quickly begin to exhibit the behaviors consistent with these labels. One can study this process among the Irish during 800 years of British conquest, among Native Americans under White domination, among Africans kidnapped into slavery, among women during the past 3000 years of Patriarchy, etc.

The same self-fulfilling prophecies often occur to immigrant groups. In America, the Irish statistically became very "lazy" when consigned to that role; but when enough individual Irish people had resisted the label strongly enough to become a powerful (and pugnacious) force in politics and business, the "laziness" of the Irish-Americans as a whole group "miraculously" began to decline. (Then it became the turn of the Puerto Ricans to play the Bottom Dog role...)

I once knew a woman in Chicago whose daughter had her whole academic life changed by labels. On entering grammar school, this girl *seemingly* scored low on an intelligence test and got assigned to a "slow learners" class. Due to the tracking system, the girl remained among the slow learners all through her eight years in that school. Then, entering high school, the girl took another intelligence test and scored in the top one per cent. She then got placed in an "accelerated" class and began showing the high intelligence which had remained dormant all through grammar school.

Did the new label create the newly apparent intelligence? Or should we just assume that the grammar school test results got mixed up and this girl received somebody else's score by mistake? I prefer the latter theory, but...on the other hand...

Every time I pass through customs between Mexico and the U.S. I feel certain sensations of anxiety. I "know" I have no illicit drugs in my car, but I begin to wonder, confronted by the hostile and suspicious eyes of the Texas Border Patrol, if some damned drug or other somehow got into the car without my knowledge... Did somebody who dislikes my books "plant" some to frame me? Did some young idiot admirer of my works slip some into a video cassette case, a book or other gift as a surprise, not knowing I intended to cross a border the next day? Do these Border people sometimes "plant" drugs themselves, to improve their arrest record? Like Joseph K. in *The Trial* I begin to feel sure they will find me guilty of something, even though I do not know of any crime I have committed.

When I finally get through customs, I feel an irrational sense of freedom, victory and personal vindication.

A friend of mine who the police once questioned in connection with a rape, went through the same sensations. At first, when the two officers began asking him to account for his movements that afternoon, he felt sure that he would get arrested, even though innocent. When the police asked several neighbors to confirm his alibi, and they did — he happened to have appeared very visible to them, working on his front lawn, when the crime occurred 20 blocks away — the police let him go at once. He felt irrationally happy, as if he had "gotten away with something." When people with guns treat you as possibly guilty, you begin to feel possibly guilty.

I suspect that African Americans and women will understand this section better than the average white male will.

Exercizes

1. The Cold War, now seemingly coming to a close, lasted 45 years (1945-1990). Discuss the role of self-fulfilling prophecies in American and Russian foreign policy during that 45 years. Discuss especially the Arms Race.

2. Children, psychologists say, tend to believe things literally. Discuss the following typical parental utterances

and how they might function as self-fulfilling prophecies; discuss also if any of you heard these in childhood and have accepted them as Life Scripts ever since.

A. You never do anything right.
B. You're so lazy you're going to end up on Welfare.
C. You have a terrible temper. You're going to hurt somebody seriously some day.
D. You've never been a healthy child.
E. Don't let me ever catch you doing that again.
F. You're going to get fat as a pig, the way you eat.
G. You're just not as smart as your brother.
H. Don't touch that part of your body again or you might go crazy.

3. Discuss the following Game Rules as self-fulfilling prophecies.

A. "The poor you will always have with you." (J. Christ)
B. "Jews make the best doctors."
C. "We will put a man on the moon within a decade." (J. Kennedy)
D. "We don't have the money to build more housing."
E. "We will build the Star Wars technology and place it in orbit, no matter how much it costs."
F. "The masses are feminine. They want a strong man to lead them." (A Hitler)
G. "All men are created equal...and endowed by their Creator with certain unalienable rights...among these rights are life, liberty and the pursuit of happiness." (T. Jefferson)
H. "Someday we can abolish hunger" (various 19th Century Futurists)
I. "We can abolish hunger by 1995." (R. Buckminster Fuller)

FIFTEEN

Psychosomatic Synergy

Let us return to psychosomatic medicine, since it illustrates the principle of the self-fulfilling prophecy in a peculiarly dramatic way.

In 1962 a young man named Vittorio Michelli arrived at the Military Hospital of Verona, Italy, suffering from a progressive carcinoma of the left hip. The whole hip appeared eaten away by the cancer and the left leg seemed about to separate from the body. Despite all the doctors could do, Michelli's disease worsened and the actual bone of the pelvis began disintegrating. The case seemed hopeless.

On May 24, 1963, Michelli left the hospital and went to Lourdes, where he bathed in the allegedly "miraculous" waters and experienced what he described as sudden sensations of heat moving through his body. His appetite, which had virtually disappeared, suddenly returned and he began eating heartily again. He felt new life and new energy; he began to gain weight. After about a month he returned to the hospital and had an X-ray examination.

The tumor appeared visibly smaller. In follow-up examinations, the doctors found that the tumor had completely disappeared and the bone began to regrow and completely reconstructed itself.

Some people (Roman Catholics, New Agers, heretical holistic physicians, etc.) will eagerly believe this yarn. Other people (the Committee for Scientific Investigation of Claims of the Paranormal (CSICOP), the American Medical

Association, old-fangled Village Atheists etc.) just as eagerly wish not to believe it at all, at all.

My source for the Michelli's case: "Healing, Remission and Miracle Cures," by Brendan O'Regan, Institute of Noetic Sciences, May 1987. O'Regan's source: the International Medical Commission on Lourdes, consisting of 25 scientists, including four physicians, four surgeons, three orthopedists, two psychiatrists, a radiologist, a neuropsychiatrist, an ophthalmologist, a pediatrician, a cardiologist, an oncologist, a neurologist, a biochemist and two general practitioners. Ten of these 25 scientists hold chairs in medical schools.

Six thousand alleged cures have been claimed at Lourdes since 1858 and only 64 have passed the scrutiny of the International Medical Commission. The Michelli case passed all their protocols of authenticity.

Since millions have visited Lourdes in the hope of such "miracles", and only 63 others beside Michelli have passed scientific investigation, these "miracles" do not seem, to me, to prove the omnipresence or the omnibenevolence of the Catholic "God". In fact, in my personal judgment, were I to accept that "God", I would wonder why the hell He only cures people who travel to Lourdes, and only a few of them, and doesn't have the compassion to cure everybody at once. Rather than assuming that a personalized anthropomorphic "God" does the things that happen at Lourdes, I prefer to consider Lourdes a trigger, and perhaps not the only trigger, that can set off a healing process in certain people prepared for such a synergetic bio-chemical-physical transformation.

In 1957, agitated by Martin Gardner and other Inquisitorial dogmatists later prominent in CSICOP, the U.S. government burned all the books of Dr. Wilhelm Reich, invaded his laboratory to smash his experimental equipment with axes, and threw Dr. Reich in jail where he shortly died of a heart attack. Dr. Reich, and about 18 other physicians working with him, had been reporting good results treating a variety of ailments with a device Dr. Reich invented, called the Orgone Energy Accumulator, which allegedly concentrated an alleged healing energy called Orgone.

All the books of Reich remained out of print in this country for over ten years[1] and two still remain out of print, even though one concerns the extremely serious health problem of atomic radiation. The journals published by Dr. Reich's Orgone Institute also remain out of print.

A few copies of Reich's journals can be found in the private libraries of various medical heretics. I looked through a pile of them once and found x-rays, taken by Dr. Victor Sobey, of a tumor that visibly shrank during a series of orgone treatments.

Members of CSICOP will insist that Dr. Sobey faked those photos, probably. Those less committed to Dogma will, I guess, have to choose between two Heresies: (1) Despite the Infallible Authority of Government Bureaucrats, the damned "orgone" does exist, after all, or (2) the belief in "orgone" can cause patients to boost their own immune systems and fight off otherwise fatal disease.

Curiously, the first perceptible effect of using an orgone accumulator, reported by all who have tried it, consists of a sensation of heat moving through the body. Just like the Michelli case at Lourdes...[2]

But let us return to Krebiozen, a matter seemingly less eldritch than cures at Lourdes or tumors destroyed by an officially non-existent energy. A very successful Krebiozen treatment, reported by Dr. Philip West, involved a cancer patient called "Mr. Wright" in Dr. West's account. Mr. Wright had fever, multiple tumors and could not get out of

---

[1] In a review of my book, *The New Inquisition,* Robert Sheaffer argues that worrying about book-burning amounts to "shedding crocodile tears" when the books only remained banned for ten years. Since most of the books banned in Nazi Germany came back into print in about ten years, presumably Mr. Sheaffer regards those who expressed horror about this as also "shedding crocodile tears." He also neglects to mention that two of Dr. Reich's books remain banned after 43 years.

[2] Sensations of heat moving through the body also appear in kundalini yoga. Funny coincidence.

bed when treatment began. The staff believed he would die in a matter of days. Within a week after Krebiozen treatment began, Mr. Wright got out of bed, began walking about the wards and chatted happily with everybody, convinced that a cure had taken place. His tumors had shrunk to half their previous size.[1]

Later, alas, when Mr. Wright learned that other patients did not respond so favorably to Krebiozen, and that doctors had begun to consider the chemical worthless against cancer, he became depressed and worried. His tumors began growing again, he returned to his bed, and he died.

Fundamentalist Materialists can only rejoice over this downbeat ending if they conveniently forget that orthodox allopathic medicine has no explanation of why the tumors measurably shrunk when Mr. Wright believed he had gotten a "miracle cure".

On the other side of the dark coin of psychosomatic synergy: a South Sea shaman points a "death bone" at a tribesman who has offended him. The victim receives the best possible medical care from sympathetic doctors, who don't believe in Black Magic, but he shortly dies anyway. It appears that the unfortunate man died of the *belief* that "death bones" can kill people.[2]

Thousands, or tens of thousands, of sick people get cured every year by Christian Science practitioners and other "faith healers". The outstanding Christian Science cures appear quarterly in *Christian Science Sentinel*. Look through a year or so of back issues and you will find a plethora of seeming cures of seemingly real cases of asthma, cancer, hypertensions, headache, sinus congestion and just about anything and everything in the catalog of human illness. The American Medical Association does not like to look at these reports at all, and CSICOP would probably like to burn them — but only for ten years, hopefully.

Without Christian Science, shamanism, orgone or anything of that sort, the Liberal activist, Norman Cousins, has

---

[1] See *The Psychobiology of Mind-Body Healing* by Ernest Lawrence Rossi, Ph.D., Norton, 1988, pages 3-8.

[2] Rossi, op. cit. p 9-12.

three times cured himself of major illnesses using the hypothesis that each human contains a healing energy that most of us do not know how to use.

Sent to a tuberculosis sanitarium at age 10, Cousins observed that the optimistic patients statistically tended to recover and get released, while the pessimists did not. He consciously became an optimist, recovered and has led a rich, productive life, editing the *Saturday Review* for many years, founding the Committee for a Sane Nuclear Policy etc.

In 1979, Cousins contracted a rare disease, *ankylosing spondylitis,* which slowly paralyzes the body and inevitably (except for Cousins) ends in death within a year. Cousins checked out of the hospital into a hotel (a safer place to live when ill, and usually much, much cheaper...) and treated himself with high doses of laughter (he spent his days looking at comedy on video, especially Marx Brothers movies and *Candid Camera).* A heretical physician also gave him massive doses of Vitamin C intravenously. Cousins recovered totally and walked normally again, the first such cure of this disease in medical literature.

In 1983, Cousins suffered a myocardial infarction and congestive heart failure — a combination that usually results in panic and death. Cousins refused to panic or die. He now teaches at the UCLA Medical School, probably the only layman on the staff, trying to show doctors how to activate this fighting spirit, or healing spirit, within each patient. (Recounted by Rossi, *op. cit.* pages 13-15, also see Cousins' own book, *Anatomy of an Illness.)*

"Spontaneous remission" — the sudden disappearance of an illness, without *any* known cause, or any contributing factor in the form of belief in Christian Science or Orgone or anything of that sort, including Mr. Cousins's healing/fighting spirit — happens so frequently that every doctor I have ever questioned on the matter admits to having seen some cases. Nobody understands "spontaneous remission" and there appears strong evidence that medical bureaucracy, as an organized political-economic entity, does not even want to think about it. Brendan

O'Regan, *op. cit.* found, after a prolonged search of medical data bases, that *only two books seem to exist in English-language medical literature about spontaneous remission, and both do not appear in print at present. You need to go to a rare-book dealer to find them.*

Dr. John Archibald Wheeler, called the father of the hydrogen bomb in some circles (others attribute paternity in that regrettable case to Dr. Edward Teller) has repeatedly urged that the simplest, most honest explanation of quantum paradoxes holds that the known universe results from the observations of those who observe it. This "observer-created universe" bears an uncanny resemblance to some of our data about "self-fulfilling prophecies," it begins to appear.

Of course, the ideal observer of quantum mechanics remains a creature hooked up to many subtle instruments. In psychology, the "observer" remains a bag of protoplasm, the resultant of genes, imprints, conditioning and learning. The genes presumably appear at random throughout the population; the imprints occur by accident at points of imprint vulnerability; conditioning and learning depend on family tradition, etc. and these factors, rather than a whimsical (or perverse) "God" probably account for who will and who will not respond to Lourdes, or to Christian Science, or to a shaman's "death-bone", etc.

Thus, in a famous Sufi story, Mullah Nasrudin, wisest man in Islam, entered England on a visit. "Do you have anything to declare?" asked the customs inspector.

"No — sssssst, bzzz — nothing at all."

"How long do you plan to stay?"

"Oh, about — sssssssszzzzt, bzzz — about three weeks."

"By the way, where did you learn English?"

"From the — bzzz, bzzz, sszzzzzzzbzz — radio."

Our brain software programs what we will see and will not see, just as the software in my computer programs what I can and cannot do with this page. (I decided to write this on Microsoft Word and find I cannot do some things I could do with MacWrite, and can do some things I could not do with MacWrite.)

But, if our brain software programs our selves and our universes, who programs our brain software? The accidents of history and environment, it seems — in most cases. But learning to internalize and use the principles of Quantum Psychology (or similar systems) adds a new factor. In that case we can gradually learn to program our programs... Dr. John Lilly calls this *metaprogramming.*

Exercizes

1. Each member of the study group had a particular gloss or reality-tunnel imposed in childhood. Discuss your parents' gloss and to what degree this still determines the universe you perceive.

2. Play-act that your group have all grown up in Moslem nation. Discuss how that would influence your reception of the ideas in this chapter.

3. Try the same exercize with the group play-acting a class of engineers in a Moscow university.

SIXTEEN

Moon of Ice

Readers of James Joyce's *Portrait of the Artist as a Young Man* will recall the horrific opening scene, in which the child, Stephen Dedalus, becomes thoroughly terrified by a superstitious servant, who tells the boy that if he does not apologize for some unspecified "sin", eagles will come and pull out his eyes. Stephen hides under the table while the threat, with its awkward and unintentional rhyme, pounds through his mind: Pull out his eyes/Apologize/Apologize/Pull out his eyes...

Joyce scholars regard this sequence as autobiographical. In an early fragment by Joyce, in the Cornell Joyce Collection, Joyce appears as the boy threatened with the eye-devouring eagles.

When Joyce began to write novels revealing the sexual side of Irish Catholic life — the Great Unspoken Secret in that country — he became the target of a campaign of vilification almost without parallel in literary history. His eyes began to bother him. He went from one eye specialist to another, and never achieved more than temporary relief. One of the eye specialists said Joyce's problem had psychological roots, but offered no suggestions about how to reach and remove those roots. Others resorted to the knife. Joyce underwent eleven painful eye operations in seventeen years, and became "legally blind" although not "medically blind" toward the end of his life.

The fact that Joyce put the eagle/eye story at the beginning of his most autobiographical novel indicates that, on some level, he understood the "curse" that had been laid on him. It appears that, like the Polynesian tribesman victimized by the death-bone, Joyce could not resist the "curse" — despite his agnosticism and skepticism. This perhaps indicates the *degree* of our malleability during those sensitive moments that ethologists call points of *imprint vulnerability*. And it may also indicate Joyce's awareness of the pain his books caused to pious Catholics. (He never did apologize...)

Kenneth Burke, who first suggested that Joyce's eye problems resulted from that traumatic early imprint, has also suggested that Darwin understood, as well as Joyce, the pain and rage his work caused ordinary Christian readers. Darwin, Burke commented, had so many inexplicable and incurable medical problems that he treated himself with higher and higher doses of opium.

An old English drinking song, allegedly humorous but also (I think) actively nefarious, begins as follows:

> Oh, my name it is Sam Hall, it is Sam Hall,
> Damn your eyes!
> My name it is Sam Hall,
> And I hate you one and all,
> Damn your eyes!
> Yes, I hate you one and all,
> You're a gang of buggers all,
> Damn your eyes, damn your eyes!

When one begins to appreciate the roles of unconscious suggestibility and self-fulfilling prophecies in human life, this song seems as funny as the latest radioactive fall-out figures.

The Whitmore and Miranda rulings of the U.S. Supreme Court — ending certain police practices once ubiquitous here and still widely practiced elsewhere — resulted from evidence that quite ordinary people, not guilty of any crimes, will in many cases confess to anything the police charge, *if they cannot communicate with attorneys or anyone else* except the police officers who are holding them. (*Isolation* begins any brain-washing process. See my *Prometheus Rising*, Falcon Press, 1983, for further details.)

In County Kerry, Ireland, in 1986, a whole family named Hayes — eight people — confessed to an act of infanticide which subsequent evidence proved conclusively they had not committed. One elderly member of the family later received a diagnosis of "senile" by physicians but the other seven had no obvious mental deficiency or mental illness. They had been held in isolation for only two days before they confessed. Ireland has no Whitmore or Miranda rulings.

Now, in thinking over these cases, do not forget that girl in Chicago (I actually knew her) who got classified as a "slow learner" (academese for mildly retarded) and acted that way for eight years before she had another test and suddenly revealed genius-level IQ...

Whether or not the physical universe deserves the label "observer-created" (as suggested by Dr. Wheeler), much of the social universe appears "observer-created." (For further data on this subject, see *The Social Creation of Reality* by sociologists Berger and Luckman and *How Real Is Real?* by psychologist Paul Watzlavick.)

As anthropologists have long noted, *every society gets a close approximation of the behaviors it expects from men and women.* We have been told over and over again that "you can't change human nature," but the study of emic realities shows, quite the contrary, that almost anything can become "human nature" if society defines it as such.

For instance, the Zuni Indian tribe in the American southwest has never had a suicide and tribal lore only recalls one murder, approximately 300 years ago. No grounds exist to believe the Zunis landed here from another planet. They appear human. They merely have a different emic reality than that of White Americans, who have very high suicide and murder rates, or Swedes, who have a very low murder rate but a comparatively high suicide rate.

As noted by Malinowski, the Trobriand Islanders never had a reported rape until after Christian missionaries brought them our Occidental reality-tunnel.

Virtually everybody believed in witches in the
Occidental world until about 200 years ago, and also believed the proper treatment for this condition consisted of
charbroiling the suspects. This idea became unfashionable
during the Age of Reason, and 20 years ago nobody would
have expected it to return. In 1990, however, a large percentage of Protestants and several police chiefs believe in a
nationwide "Satanic" underground and, although nobody
has yet gone to the stake, a new witch-hunt obviously has
swept our country.

The Nazis believed the moon consisted of solid ice. Brad
Linaweaver's superb science-fiction novel, *Moon of Ice*,
concerns a parallel universe where World War II ended in
a truce, rather than total victory for the allies. In Nazi
Europe, the "moon of ice" theory still reigns supreme in
government-run Universities, learned societies, etc. while
in anarchist America (in that universe, we became pacifist,
isolationist and finally anarchist) the orthodox model of
the moon remains dominant. When the Nazis land a
spaceship on the moon and find no ice, all the data of the
flight becomes Top Secret and the Europeans never learn
of it.

Does this plot seem unlikely to you? Look back a few
pages and see what happened to the only two studies of
spontaneous remission written in English.

Exercize

Obtain a copy of *High Weirdness by Mail*, by Rev. Ivan
Stang, (Simon and Shuster, 1988), a catalog of dissident
groups in the U.S., covering the full spectrum from the
fairly plausible (and possibly important) to those that
appear totally "nutty" to almost all the rest of us. Pick out
five groups that seem sane and plausible, and five that
seem totally crazy, and send for a packet of their literature.
(Stang gives mailing addresses for all groups he reviews.)
Study the literature and discuss in the group.

Do some of the plausible groups look less plausible
when analyzed operationally and skeptically? Do some of

them seem, maybe, as important as a dissident group in Nazi Europe publishing evidence that the moon doesn't consist of ice? And — do any of the "nut groups" look less nutty when you analyze their arguments?

SEVENTEEN

Taking the Mystery Out of "Miracles"

If one society has no rapes, and another has no suicides, emic reality (brain software) programs etic reality (what happens to people within an emic reality) more than we generally realize.

The seriousness of this issue as a philosophical problem seems immediately evident. Even if "meaningless" technically, the New Age bromide, "You create your own reality" has a kind of connection to the actual facts. Society seems to create a reality-tunnel, which each member modifies to some extent, and conflict grows out of the delusion that "I have the one correct reality-tunnel" when "I" have to deal with somebody who has another "one correct" reality-tunnel.

The seriousness of this issue as a psychological/sociological factor seems even more staggering than its philosophical implications. We cannot say exactly what would happen if, e.g., George Bush or Mr. Gorbachev accepted Buckminster Fuller's claim that we can abolish starvation by 1995, but something very dramatic and surprising would undoubtedly happen, something that would alter our notions of "inevitability."

Since body-changes seem more "miraculous" than social changes, let us look further into the matter of psychosomatic synergy.

The extent to which the health/illness spectrum deserves the label "observer-created" and/or "self-fulfilling prophecy" appears strikingly in studies of placebo effectiveness.

According to several careful double-blind studies cited by the invaluable Dr. Rossi (*op. cit.* pages 15-19):

Placebos proved 56 per cent as effective as morphine in six double-blind studies;
Placebos proved 54 per cent as effective as aspirin in nine double-blind studies;
Placebos proved 56 per cent as effective as codeine in three double-blind studies.

("Effectiveness" rating in these studies indicates how much relief from pain the patient reported.)

In other words, slightly more than half the time a patient obtains as much benefit from the belief that she or he received a pain-killer as they would obtain from actually receiving a pain-killer.

As O'Regan points out, *op. cit.* we now have reason to believe that almost all medical treatment throughout almost all history worked on placebo principles. In other words, modern biochemistry indicates that before the discovery of antibiotics, (i.e., *before the 1930s*) virtually all the medicines used by doctors had no actual effectiveness. The patients got better, when they did, because the doctors believed in their useless potions and the patients acquired the "faith" from the doctors.

This last paragraph has more than historical interest. According to the Office of Technology Assessment, *only 20 percent of established medical procedure in the U.S. today has been validated in rigorous randomized, double-blind, placebo-controlled studies.*[1] Thus, 80 percent of what our doctors do rests simply on precedent and high hopes. Since more than 20 percent of us survive American medicine, a great many placebo cures must still occur daily, as they did before the 1930s.

My own books, especially *Prometheus Rising,* give numerous examples of how optimism (a "Winner Script" in the language of Transactional Analysis) can resolve psychological and social problems that seem incurable to

---

[1] *Los Angeles Weekly,* Sept. 16-22, 1988. Article, "Blinded by Science?" by Carolyn Reuben.

those obsessed by pessimism (a "Loser Script" in T.A.) But most people in our society — and even some scientists — still feel that something "miraculous" has occurred if a Winner Script can conquer not just negative feelings and bad social adjustment but physical cancers and other "no-nonsense" body diseases, or if a Loser Script can cause a person to lie down and die like the victims of "death bones".

As I have already indicated, this sense of "miracle" and "mystery" derives from our traditional dichotomy of "mind" and "body" and our habit of thinking that any-thing we have split verbally must reflect a similar Iron Curtain in the non-verbal existential world. (Similarly, physicists, having traditionally split "space" and "time" found themselves confronted by terrible mysteries and confusions at the end of the 19th century when this map clearly no longer fit the territory. It took the genius of Einstein to re-unify verbally what had always remained unified non-verbally: he gave up "space" and "time" and wrote of space-time, and the mysteries quickly found solutions.)

As suggested earlier, the mystery of "mind" and "body" begins to disappear if we speak and think without this dichotomy, in terms of the organism-as-a-whole. To quote Bowers:

> The tendency to split etiological factors of disease into psychic or somatic components, though heuristic for many purposes, nevertheless perpetuates, at least implicitly, a mind-body dualism that has defied rational solution for centuries. Perhaps what we need is a new formulation of this ancient problem, one that does not propose a formidable gap between the separate "realities" of mind and body...
>
> If information processing and transmission is common to both psyche and soma, the mind-body problem might be reformulated as follows: How is information, received and processed at a semantic level, transduced into infor-mation that can be received and processed at a somatic level, and *vice-versa*? That sounds like a question that can

be more sensibly addressed than the one it is meant to replace.[1]

Transduction, in Information Theory, designates the translation of form from one information system to another. For instance, if I speak to you on the phone, the transmitter transduces my words (sound waves) into electrical charges which — if the phone company does not screw up again — travel to the receiver in your hand, where they become transduced again, back into sound waves, which you decode as words.

Similarly, as I sit here at my MacPlus, the keys I hit look like letters of the English alphabet but hitting the keys creates binary (on-off) signals stored in computer memory. The words have become transduced into electrical charges. When I print this up — if the computer does not screw up again — these electrical charges will become transduced once more, into words that you can read.

The "mind-body problem" has "defied rational solution," I suggest, because any question asked within that framework qualifies as totally "meaningless" in the Copenhagen and logical positivist sense. However, if we drop "mind and body" from our vocabulary and replace them with "psychosomatic unity" or "psychosomatic synergy" — as physicists after Einstein dropped "space" and "time" and replaced them with "space-time" — we begin to approach an area where meaningful questions and answers can appear. But this requires, as Bowers suggested, formulations in terms of Information Theory and transduction.

A strong set of negative beliefs (a Loser Script) appears, to neuroscience, as an imprinted and/or conditioned and/or learned network of bio-chemical reflexes in the cortex of the brain. Since communication exists between parts of the brain, and between the brain and other body systems, these "negative beliefs" can easily transduce into bio-chemical reflexes of the organism-as-a-whole. Specifically,

---

[1] K. Bowers, "Hypnosis: An Informational Approach," *Annals of the New York Academy of Sciences,* 296, 222-237 (1977)

the "belief" reflexes in the cortex get transduced into neurochemical and hormonal processes when they pass through the hypothalamus, an ancient little part of the back brain which regulates and/or influences many body programs, *including the immune system.*

Among the chemical systems regulated by the hypothalamus and transduced to the immune system we find a variety of *neuropeptides*, including the now-famous *endorphins*, which act as tranquilizers and pain-killers quite similar to opium.

Neuropeptides have a curious duality about them which reminds me of the photons (and electrons) of quantum mechanics. Those quantum entities (or models?), you remember by now, sometimes act as waves and sometimes as particles. Similarly, neuropeptides sometimes act as hormones (chemicals causing changes in body function) and sometimes as neurotransmitters (chemicals causing changes in brain function).

None of these little blighters ever heard of Aristotelian logic, I guess.

Acting as neurotransmitters in the brain, neuropeptides perform many interesting known functions (and probably many not yet known...) Most significantly, they allow the opening and perhaps the imprinting of new neural pathways and "networks" and/or "reflexes". This means that a heavy dose of new neuropeptides in your brain, just like a dose of LSD or some other psychedelic, will cause you to perceive and "think" (organize and interpret perceptions) in new and original ways — to drop your familiar gloss and "see" through other glosses...to leave your rigid reality-tunnel and enter a multi-choice reality-labyrinth... to transcend *modeltheism* (dogma) and spontaneously feel-think in the manner of the "model agnosticism" of post-Copenhagen physics...

Whatever metaphor from the behavioral sciences one uses, the process means, in ordinary terms, decreased rigidity, increased creativity; less compulsion, more sense of choice.

In terms of Information Theory, this appears as a dramatic increase in the amount of information processed per second. The more new circuits opened in the brain, the more new information you notice in even the simplest and most familiar objects or events. To quote Blake, "The fool sees not the same tree that the wise man sees."

A really massive rush of neuropeptides, then, will subjectively appear as "rebirth" or "seeing a whole new universe" or a transcendence of what had seemed ineluctable limitations. Many will describe this in religious metaphors and say "the Spirit overcame me," etc. Blake speaks of seeing "infinity in a grain of sand."

When they leave the brain and begin acting as hormones in the body, neuropeptides interact with all significant systems, including the immunological system. Increased neuropeptide activity, therefore, correlates with increased "resistance" to disease, an inner sense of "feeling better" and the kind of upsurge of hope that propelled Mr. Wright out of bed and set him walking about the wards chatting happily with everybody.

A few miscellaneous observations — taken, like most of the above, from Rossi, *op. cit.* — will illustrate these synergetic relationships somewhat further.

1. Those who respond best to placebos also register high on awareness of synchronicities. Since synchronicity only "makes sense" in a holistic or synergetic model of the universe (and appears "nonsensical" or "impossible" in a mechanistic model) such people already have an intuitive sense of holism, which makes it easier for them to "allow" holistic processes to occur in brain/body systems.

2. Those who respond least to placebos not only deny synchronicity but appear "rigid and stereotypical" in their thinking. Thus, placebos would probably not work for members of CSICOP. It almost appears that some people would rather die than allow a cure that looks to them like "magic".

3. Memory now appears mood-dependent. When we feel happy, we sincerely remember our lives as generally happy; when we feel said, we conversely remember our

lives as total disasters, etc. The "observer" who creates our experienced universes not only appears unaware of its own creativity, but re-edits everything in terms of current mood (i.e., current neurochemical activity in the brain).

4. Many studies indicate that the neuropeptide activity in the brain — reassociating, or re-glossing, or moving from a rigid reality-tunnel to a multi-choice reality labyrinth — seems as important in healing as the chemical boost that neuropeptides give to the immune system. In other words, as our ability to process more and more information increases, our resistance to unwellness (in general) also increases.

A world of many options never "feels" as dreadful as a deterministic or mechanical world.

5. The brain never "remembers" like a tape recorder or repeats like a parrot. *Even the most rigid and compulsive types* (Catholics, Marxists, members of CSICOP, etc.) *do a lot more re-associating, re-framing and creative editing than they consciously realize.*

Dr. Rossi summarizes the current evidence by saying bluntly that what judges tell witnesses to do in a court-room — tell one version of experience, and stick to it, without re-editing — seems unnatural and nearly impossible for the human brain. It seems we never do that, exactly.

At best, we may convince ourselves and others that we have done it, for a short period. The attorney for the opposition can usually tear that charade apart — to the utter consternation of witnesses who have never heard of Transactional psychology or quantum logic and still believe in the Aristotelian/medieval "one objective reality."

6. Beta wave activity in the brain correlates with outer-directed activity and dominance of sympathetic nervous system functions. Alpha wave activity, and lower brain frequencies, correlate with inner-directed passivity and dominance of parasympathetic nervous system functions.

Daily practice of yoga, which often correlates with improved health, decreases the total amount of beta activity/outer directed attention/sympathetic nervous system function and conversely increases alpha or theta wave activity/inner directed attention/parasympathetic nervous system functions.

Hypnosis, whatever positive suggestions it may implant in the cortex to transduce into neurochemical immunological boosts, also begins with telling the patient to close her or his eyes and become more relaxed. Both closing the eyes and relaxing move the patient from beta waves/outer attention/sympathetic system to alpha-theta waves/inner attention/parasympathetic system.

(The controversial Dr. Reich, incidentally, used muscular relaxation techniques to move patients from sympathetic nervous system dominance to increased parasympathetic nervous system activity. But since Certified Government Bureaucrats condemned his ideas and burned his books, you all know he "was really" a nut, right?)

7. Since the neuropeptides travel through virtually all the body fluids (blood, lymph, cerebrospinal fluid etc.) as well as between neurons, the neuropeptide system acts more slowly but more holistically than the central nervous system.

The experimental attitude differs totally from "common sense" — the former accepts that we may, at any time, discover new information that will profoundly alter our model of the universe, while the latter assumes we know the basic truth already and, at worst, will only have to modify it slightly when new data appears. Thus, I do not mind confessing that I have tried faith healers on occasion, experimentally. I have enough conservatism (or cowardice) to have tried these experiments only with minor ailments that did not seem likely to become serious hazards to my survival.

The results exactly conformed to point 7 above, even though I knew nothing about neuropeptide activity until a few years ago. With each "healer", I felt nothing very dramatic during the treatment, and each time I left with a sense of disappointment and increased skepticism (about that school of faith healing, or that healer). In a few hours, I began to notice a slight decrease in symptoms and a upsurge in "new energy." Within a day, all symptoms disappeared and my health again appeared normal. I did not know how to explain this effect until I read about the slow-motion holistic activity of neuropeptides.

Perhaps, even after these neurochemical functions seem clear to the reader, an aura of "spookiness" still lingers about the whole subject. Let us look at the belief/neuropeptide/immunological loop in slow motion, then. This will perhaps appear less "spooky."

According to *Brain/Mind Bulletin* (May 1988) John Barefoot of Duke University has found a negative correlation between suspiciousness and longevity. In a sample of 500 older men and women whose health he monitored for 15 years, Barefoot discovered that:

(a) those who scored high on suspiciousness, cynicism and hostility died sooner than all others;

(b) this high mortality among those with Loser Scripts remained constant when compared by age, by sex, by previous health, by diet and even by "bad habits." (Those who smoked and remained generally optimistic lived longer than those who smoked and worried about it.)

(c) those who scored highest on hostility had a death rate more than six times higher than others.

In a related study (*Brain/Mind Bulletin* August 1988) Shelley Taylor of UCLA and Jonathan Brown of SMU refuted the conventional idea that those who score high on "mental health" have fewer illusions than others.

Quite the reverse, according to this study: those who score high on "mental health" generally have a number of illusory beliefs. Among the most common illusions of the mentally healthy:

(a) overly positive views of themselves;
(b) convenient "forgetting" of negative facts about
     themselves;
(c) illusory beliefs about having more control than they
     do have;
(d) "unrealistic" optimism about themselves;

(e) "unrealistic" optimism about the future in general;
(f) "abnormal" cheerfulness.

Would you want to have those kinds of "illusions" or would you rather stick to "hard realism" and die sooner than those deluded fools?

In closing this chapter, I would like to give another case history and a bit of self-revelation. At the age of 2 years, in 1934, I contracted polio — a rather widespread disease among children up until the Salk vaccine. Dr. Salk had not discovered the vaccine yet, in 1934, and the medical prognosis held that I would never walk again.

My parents eventually found a doctor who had decided to experimentally treat some polio patients with the "Heretical" methods of Sister Kenny, an Australian nurse who had been roundly Damned and Anathematized by the A.M.A. bureaucracy. On all sides, Americans received the message that the Sister Kenny methods did not work and that they consisted of "quackery" and "witchcraft."

The Kenny system consisted of (a) something a bit like faith healing, (b) muscle massage and (c) long soaking in hot tubs.

The "faith healing" side of Sister Kenny's technique involved flat denial of the A.M.A. dogma that those crippled by polio could never walk again. The muscle massage had a lot in common with the techniques of the infamous Dr. Reich, Damned and Anathematized in the 1950s by the same medical bureaucracy that condemned Sister Kenny in the 1930s. The hot tub idea had been popular in the 19th Century, remained popular in Europe and has become popular again in California. I have no idea whether my cure resulted from one of these factors, from two, or from all three in synergy. Empirically, I recovered and started walking again. I walk normally today, with only an occasional limp when very tired, and some pedal myoclonism at night. Most people do not guess that I spent two years as a cripple.

Those treated by orthodox A.M.A. methods in those years seem to have remained in wheel-chairs for the rest of their lives.

In retrospect, I wonder how much of the condemnation of Sister Kenny resulted from the facts that she did not possess male genitalia or a medical degree (i.e., in the minds of most physicians, she "was" "only" a woman and "only" a nurse...)

I suspect that some long-term effects of the Sister Kenny treatment linger with me. E.g., I have enjoyed better-than-average health for the rest of my life, I retain a deep suspicion of all "Authorities" and Authoritarians (as you might have noticed), and I have never had the fashionable pessimism and *bon ton* despair necessary to get myself included among Serious Novelists in the judgment of New York critics. Like the people in the Taylor-Brown study, I seem "unrealistically" optimistic about myself and the future and "abnormally" cheerful. This annoys quite a few people.

Exercizes

1. Buy one of the hypno-tapes on how to build self-confidence and increase self-healing now widely advertised. (I especially recommend those available from Acoustic Brain Research, 640 Ocean View Drive, Friday Harbor, WA 98250, U.S.A.) Play this tape at every weekly group meeting. Observe any changes in the behavior and/or the general health of the members of the group.

2. Rewrite the following sentences in E-Prime:
A. Dr. Reich was a quack.
B. Sister Kenny was a quack.
C. "Everybody is a bit queer except me and thee, and sometimes I wonder about thee."
D. Cancer is caused by worry and depression.
E. Cancer is caused by a virus.
F. The cause of schizophrenia is sexual repression.
G. Schizophrenia is caused by genetic predisposition.
H. She is a Catholic, so therefore she is against abortion.
J. "Evolution is no longer a theory; it is a proven fact." (Variations on this came forth from several

biologists during recent controversies with Bible Fundamentalists.)
K. "The whole New Age is Satanic." (Rev. Pat Robertson)
L. "Reality is whatever you think it is."
M. "Nothing is. Nothing becomes. Nothing is not." (Aleister Crowley, *Book of Lies (falsely so called)*)
N. "Bob is. Bob becomes. Bob is not. Therefore, Bob is nothing." (Ivan Stang, *Book of the Sub-Genius*.)
3. Rewrite the following *questions* in E-Prime.
A. Are all diseases psychosomatic?
B. Are some UFOs really alien spaceships?
C. What is Justice?
D. What is Art?
E. What is the cause of poverty?
F. What is the cause of war?
G. Why are there so many homeless people in this rich country?
H. "It's pretty, but is it Art?" (Kipling)

PART FOUR

Schroedinger's Cat & Einstein's Mouse

"Art imitates nature."
— Aristotle

"Nature imitates art."
— Oscar Wilde

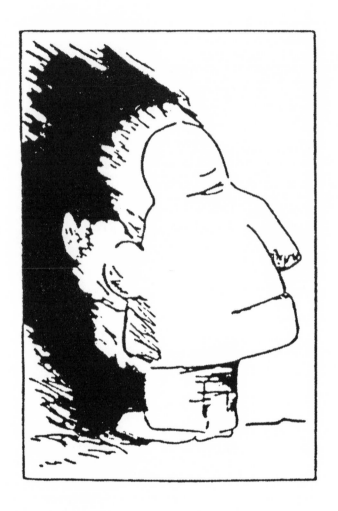

You can see the above illustration two different ways. Can you see it both ways at the same time, or can you only change your mental focus rapidly and see it first one way and then the other way, in alteration?

"The true essence of things is a profound illusion."
— F. W. Nietzsche

EIGHTEEN

Multiple Selves & Information Systems

Between 1910 and 1939, Charlie Chaplin always played the same character in all his films — the beloved little Tramp that became world-famous. In 1939, Chaplin wrote, directed and starred in *The Great Dictator*, in which the little Tramp did not appear. Instead, Chaplin played *two* characters — a tyrant, based on Hitler, and a Jewish tailor, one of Hitler's victims. Audiences all over the world (except Germany, where the authorities banned the film) complained, mournfully and angrily, that they missed the little Tramp. Chaplin, however, having gotten rid of the Tramp once, never did bring that persona back. In later films, he played many characters (a serial killer, a kindly old vaudevillian, a deposed king), but never the Tramp. People still complained that they wanted to see the Tramp again, but Chaplin went on creating new characters. (We will leave it to Jungians to explain why Chaplin had to become two opposite characters before he could personally escape the Archetype of the Tramp...)

Many actors have had equally hard battles in getting detached from, if not a specific character, a specific type. Humphrey Bogart remained stuck in villain roles, usually gangsters, for nearly a decade before he got to play his first hero. Cary Grant never did escape from the hero type — either the romantic hero or the comic hero; when Alfred Hitchcock persuaded him to play a murderer, in *Suspicion,* the studio over-ruled both of them and tacked on a surprise ending in which the Grant character did not commit the murder, after all. Etc.

Back in "the real world," if a member of a family changes suddenly, the whole family suddenly appears agitated and disturbed. Family counselors have learned to expect this, even when the change consists of something everybody considers desirable — e.g., an alcoholic who suddenly stops drinking can "destabilize" the family to the extent that another member becomes clinically depressed, or develops psychosomatic symptoms, or even starts drinking heavily (as if the family "needed" an alcoholic).

It seems that we not only speak and think in sentences like "John is an old grouch" but become disoriented and frightened if John suddenly starts acting friendly and generous. (Audiences rejected the previously "lovable" Chaplin most vehemently when he played the multiple wife-killer in *Monsieur Verdoux.* Probably, audiences would not have felt upset if the role had gone to the actor who originally wrote it for himself and sold it to Chaplin when the Hollywood moguls blacklisted him — Orson Welles.)

If Dickens' Scrooge had changed, in actuality, as he changed in the book, several people in his social field would have suddenly developed bizarre behaviors they had never shown before...

Chaplin, amusingly, once made a comedy about the chaos created by a man who conspicuously does not exhibit the "isness" or "essence" our subject-predicate language programs us to expect, *City Lights.* In this film, the little Tramp encounters a millionaire with two entirely different personalities: a generous and compassionate drunk, and a greedy, somewhat paranoid sober man. The Tramp and all the other characters soon exhibit behaviors that would look like clinical insanity to the audience, if we did not know the secret none of the characters guess: namely that each "personality" in the rich man appears when brain chemistry changes.

The Russian mystic Gurdjieff claimed that we all contain multiple personalities. Many researchers in psychology and neuroscience now share that startling view. As Gurdjieff indicated, the "I" who toils at a job does not seem the same "I" who makes love with joy and passion, and the third "I" who occasionally gets angry for no evident reason seems a third personality, etc. There does not appear anything metaphysical about this; it even appears, measurably, on electroencephalograms. Dr. Frank Putnam of the National Institute of Health found that extreme cases of multiple personality — the only ones that orthodox psychiatry recognizes — show quite distinct brain waves for each "personality" almost as if the researchers had taken the electrodes off of one subject and attached them to another. (O'Regan. *op. cit.*)

Dr. Rossi defines these separate personalities as *"state specific information systems."* Not only do we show different personalities when drunk and when sober, like Chaplin's emblematic millionaire, but we have different information banks ("memories") in these states. Thus, most people have noted that something that happened to them while drunk appears totally forgotten, until they get intoxicated again, and then the memory "miraculously" re-appears. This observation of state-specific information occurs even more frequently with LSD; nobody really remembers the richness of an LSD voyage until they take another dose.

Emotional states seem part of a circular-causal loop with brain chemistry — it seems impossible, for science in 1990, to say that one part of the circle "causes" the other parts. Thus, we can now understand a phenomenon mentioned earlier, namely that we tend to remember happy experiences when happy and sad experiences when sad.

The separate "personalities" or information systems within a typical human seem to fall into four main groups, with four additional groups appearing only in minorities who have engaged in one form or another of neurological self-research (metaprogramming).

1. The Oral Bio-Survival System. This seems to contain imprints and conditioning dating from early infancy, with subsequent learning built upon that foundation.

If you stop and think about, you know how a carpet tastes, how the leg of a chair tastes, etc. You may even remember how the dirt in a flower pot tastes. This knowledge dates from the oral stage of infancy in which we take nourishment (bio-survival) through the mother's nipples and also judge other objects by putting them in our mouths. A large part of parenting an infant consists in following the little darling around and shouting "Don't put that in your mouth" whenever they try to taste/test something toxic.

Dating from Adorno in the 1940s, psychologists who do surveys on large groups (e.g., entering college freshpersons) have repeatedly noted a correlation between dislike of "foreign" and "exotic" foods and the "fascist" personality. A total *Gestalt* seems to exist — a behavioral/conceptual cluster of dislike of new food-dislike of "radical" ideas/racism/nationalism/sexism/xenophobia/conservatism/phobic and/or compulsive behaviors-fascist ideologies. This cluster makes up the well-known F-Scale (F for Fascism). Where more than two of these traits appear, the probabilities indicate that most of the others will appear.

This seems to result from a *neophobic imprint in the bio-survival system.* Those with this imprint feel increasingly insecure as they move in space-time away from Mommy and "home-cooked meals." Conversely, those who like to experiment with strange and exotic foods seem to have a *neophilic imprint* and want to explore the world in many dimensions — traveling, moving from one city or country to another, studying new subjects, "playing" with ideas rather than holding rigidly to one static model of the universe.

On this baby-level of the brain, some seem to have an imprint that clings to the familiar ("Oh, Mommy, take me home"), some have the opposite imprint that seeks novelty and exploration ("Let's see what's on the other side of the mountain") and most, following the Bell-shaped curve, have an imprint somewhere between these extremes — "conservative" on some issues, innovative on others.

Subsequent learning will tend to get processed through these imprints, and those with strong neophobic reflexes will usually, if they ever reject the initial dogmatic family reality-tunnel, settle at once into an equally dogmatic new reality-tunnel. E.g., if raised Catholic, they seldom become agnostics or zetetics; rather, they will move, like iron filings drawn by a magnet, to dogmatic atheism or even a crusading atheist "religion" like Marxism, Objectivism or CSICOP.

Since the mechanical bio-chemical reflexes on this level remain "invisible" (and cannot even reach translation onto the verbal level except in an altered state of consciousness, such as hypnosis, or under certain drugs), this hard-wired infantile information system controls all later information systems (or "selves") without the knowledge of the conscious ego.

In most cases, the "happiest" or most tranquil areas of the infantile bio-survival system — those imprinted by the Safe Space around Mommy — can only be "remembered" or re-experienced with drugs that trigger neurotransmitters similar to those activated during breast-feeding. The attempt to re-capture that state may lead to re-imprinting via yoga or martial arts, or to a search for chemical analogs, which will eventually lead to the opiates. "Disturbed" or "unhappy" (ego dystonic) imprints here may account for opiate addictions.

This oral bio-survival system makes a feedback loop from mouth to hypothalamus to neuropeptide system to lymph and blood etc. to immunological system. What Transactional Analysis calls the Wooden Leg Game — evasion of adult responsibility through chronic illness — does not appear conscious in most cases. Rather a Loser Script in this system depresses the sub-systems, including the immunological system, and renders the subject, or victim, statistically prone to more illness than average. Similarly, a Winner Script on this circuit contributes to longevity and may account for cases like Bertrand Russell (still writing philosophy and polemic at 99), George Burns (busy with three careers until 100) etc.

2. The Anal Territorial System. Since all mammals mark their territories with excretions, the "toddler" stage of development and associated toilet training produces a system of synergetic imprints and conditioning concerned with territory and what Freudians call "anality" (sadomasochism).

Those who take a Dominant imprint in this system seek power all their lives; those with a Submissive imprint seek Dominant types to lead them (the Reichian *Fuhrerprinzip*) and most people settle somewhere between these extremes, taking a masochist stance toward those "above" them (government, landlords etc.) and a sadist stance toward selected victims defined as "below" them (wives, children, "inferior races," people on Welfare, etc.)

The "self" or information system on this toddler level may function as the predominant self or "normal" personality in those whose lives center around power or it may remain "latent" usually and only emerge in conflict situations. Usually, it emerges full-blown when enough alcohol enters the brain and alters habitual circuitry. The anal-sadist vocabulary of the typical drunk ("Oh, yeah? Stick it up your ass," "You dumb ass-hole," "Up yours, buddy," etc.) recapitulates toilet training and mammalian habits of using excretions as territorial fight-or-flight signals.

People say later "He was acting like a two-year-old" or more simply "He just wasn't himself last night." These remarks signify that the toddler information system — i.e., the mammalian anal-territorial circuits — temporarily took control of the brain.

Politicians have great skill in activating this system and easily persuade large crowds to behave like small children having temper tantrums. The favorite activating device (dramatized by Shakespeare in *Henry V)* invokes *mammalian pack-solidarity* by attacking a rival pack. George Bush, perceived as a "wimp" by many, raised his popularity to unprecedented heights, just as I looked about for a contemporary illustration of this point. Mr. Bush simply invaded a small, Third World country (Panama) where a quick, easy victory came within a week. The "wimp" image vanished overnight. Any alpha male in any gorilla or chimpanzee pack, feeling his authority slipping, would have followed the same course.

This system makes a feedback loop between muscles, adrenaline, the thalamus of the brain, the anus and the larynx. Swelling the body and using the larynx to howl (muscle-flexing and noise) makes up the usual Domination signal among birds, reptiles, mammals and politicians. Study the speeches of Hitler and Ronald Reagan for further details, or just watch two ducks disputing territory in a pond.

Conversely, shrinking the body and muttering (or becoming totally silent) make up the usual Submission reflex. "Crawling away with its tail between its legs," the dog's submission reflex, does not differ much from the body-language of an employee who made the mistake of disagreeing with the boss and received a Dominator (flexing/howling) signal in response.

The ego — or self — defined by this system appears more mammalian and evolutionarily advanced than the quick reptilian reflexes of the self operating on the oral bio-survival system. Nonetheless, the personality shrinks back to the primitive bio-survival self whenever real danger appears — whenever confronted by threat to life, rather than mere threat to status. This difference between mammalian strategy and reptilian reflex explains why *there seems more "time" in the anal territorial system than in the oral bio-survival system.* In the later, mammalian system, one explores relative power signals slowly; in the earlier, reptilian system, one attacks or flees instantly.

3. The Semantic Time-Binding System. After the growing child acquires language — i.e., learns that the flux of experience has had labels and indexes assigned to it by the tribal game-rules — a new information system becomes imprinted and conditioned, and this system can continue growing and learning for a lifetime.

This system allows me to receive signals sent 2500 years ago by persons such as Socrates and Confucius. It allows me to send signals which, if I have more luck than most writers, will still find their way to new receivers 2500 years in the future. This time-binding function of symbolism gives humans problem-solving capacities impossible to most other animals (except, perhaps, cetaceans) and also allows us to create and suffer from "problems" that do not exist at all, except on the linguistic level.

With human symbolism we can produce (or learn from their producers) mathematical systems that allow us to predict the behavior of physical systems long before we had the instruments to measure those systems (as Einstein predicted that clocks in outer space would measure time differently than clocks on our planet face). We can even build complex machines that work — most of the time.

With symbolism we can also write messages so profound that nobody fully understands them but almost everybody agrees they say something important (e.g., Beethoven's Ninth Symphony).

And with symbolism we can create meaningless metaphysics and Strange Loops so weird that society grows alarmed and either locks us up or insists on "medicating" us. With such weird symbols, if not locked up or medicated, we can even persuade multitudes to believe in our gibberish and execute 6,000,000 scapegoats (the Hitler case), line up to drink cyanide cocktails (the Jim Jones case), or perform virtually any idiocy or lunacy imaginable.

If the imprints in the first two information systems differentiate us into large groups — conservatives and pioneers, dominators and followers, etc. — the semantic system allows us to differentiate ourselves still further, giving humanity more tribal eccentrics, both benevolent and malign, than any other class of animals.

We do not all live in the same universe. Millions live in a Moslem universe and find it very hard to understand persons living in a Christian universe. Millions of others live in a Marxist universe. Most Americans seem quite happy in a mixed 19th Century Capitalist and 13th Century Christian universe, but the literary intelligentsia lives in an early 20th Century Freudian/Marxist universe, and a few well-informed scientists evidently actually live in a 1997 universe. *Etc.*

The elaboration of such emic realities or reality-tunnels can reach extremes of creativity, in which a person "invents" a totally new and individualized gloss on the whole of existence. Such great creators will either win Nobel prizes (for art or science) or will get thrown in "mental hospitals," depending on how much skill they have at selling their new vision to others. Some will even get locked up in nut-houses and later become recognized as great scientific pioneers — e.g., Semmelweiss, the first physician to suggest that surgeons should wash their hands before operating.

(Ezra Pound had the peculiar distinction of winning an award from the Library of Congress for writing the best poem of the year, in 1948, while government psychiatrists insisted he "was" insane.)

The semantic time-binding system makes a feedback loop between the verbal left brain hemisphere, the larynx, the right hand (which manipulates the world and checks the accuracy of maps or glosses) and the eyes (which read words and also scan the environment).

The self existing in this system has more "time" than the self on the mammalian territorial system or the reptilian survival system. Indeed, it can speculate about "time", or about other words, and invent philosophies about timeless universes, three-dimensional time (Ouspensky), infinite time dimensions (Dunne) etc. It can invent new Gestalts which make quantum jumps in our social information banks and it can wallow in utter nonsense endlessly.

A "clever" imprint in this system usually lasts for life, as does a "dumb" imprint. Subsequent conditioning and learning all occur with the parameters of a fluent (well-spoken, clear-thinking) self or a dull (inarticulate, "unthinking") self.

4. The Socio-Sexual System. At puberty, the DNA unleashes messenger RNA molecules which notify all subsystems that mating time has arrived. The body metamorphizes totally, and the nervous system ("mind") changes in the process. A new "self" appears.

As usual, imprinting and genetics play a major role, with conditioning and learning modifying but seldom radically altering genetic-imprinted imperatives. If the environment provides a sex-positive imprint, adult sexuality will have a joyous and even "transcendental" quality;. if the environment provides a sex-negative imprint, sexuality will remain disturbed or problematical for life.

The socio-sexual system feedbacks run from front brain through hormonal and neuropeptide systems to genitalia to breasts and arms (hugging, cuddling, fucking circuitry). A "good" sexual imprint creates the archetypal "bright eyes and bushy tails," while a "bad" imprint creates a tense (muscularly armored) and zombie-like appearance.

The self or ego in this system easily learns adult Game Rules (civilized norms, "ethics"), if the sexual imprint has not had strong negative components. Where the imprint does have negative or "kinky" components, adult Game Rules do not set in place and either an "outlaw" personality crystallizes (the rapist/criminal with the archetypal "Born to Lose" tattoo) or else the Jekyll-Hyde dualism appears, well illustrated recently by several sex-negative TV preachers who got caught in some very kinky private sex-games.

Whatever system dominates at a given time appears as the ego or self at that time, in two senses:

1. People who meet Mr. A when he has the Oral Submissive self predominant, will remember him as "that sort of person." People who meet him when he has the Semantic/rational self predominant remember him as another sort of person. Etc.

2. Due to state-specific information, as discussed earlier, when you have one of these selves predominant, you "forget" the other selves to a surprising extent and act as if the brain only had access to the information banks of the presently predominant self. E.g., when frightened into infantile Oral states, you may actually think "I am always a weakling," quite forgetting the times when your Anal Dominator self was in charge, or the Semantic or Sexual imprints were governing the brain, etc.

(This analysis owes a great deal to Dr. Timothy Leary's *Info-Psychology*, Falcon Press, 1988. A discussion at greater length, less technical than Leary's, appears in my *Prometheus Rising, op. cit.*)

But, if we have a variety of potential selves rather than the one block-like "essential self" of Aristotelian philosophy, and, if each self acts as an observer who creates a reality-tunnel which appears as a whole universe (to those unaware of Transactional and Quantum psychology), then:

Each time an internal or external trigger causes us to quantum jump from one "self" to another, the whole world around us appears to change also.

This explains why Mary may say, and honestly believe, "Everybody bullies me" one day and then say, and honestly believe, "Everybody likes me and helps me" on another day, why John may feel "Everybody is a bastard" one hour and "I feel sorry for everybody; they're all suffering" the next hour.

Every person lives in different *umwelt* (emic reality) but every self within a person also lives in a different reality-tunnel.

The number of universes perceived by human beings does not equal the population of the planet, but several times the population of the planet. It thus appears some sort of miracle that we sometimes find it possible to communicate with each other at all, at all.

Quantum mechanics says an electron has a different "essence" every time we measure it (or, more clearly, it has no "essence" at all). Neuroscience reveals, similarly, that the Mary we meet on Tuesday may have a different "self" than the Mary we met Monday (or, as the Buddhists said long before neuroscience, Mary has no "essence" at all).

As we said at the beginning, the bedrock claim of existentialism holds that "existence precedes essence," or we have no "essence". Like electrons, we jump from one information system to another, and only those who have not looked closely believe that one "essence" remains constant through all transformations.

Exercizes

1. J. Edgar Hoover, head of our secret police for over 50 years, now appears to have lived the life of an active homosexual. He kept files on the sexual behavior of politicians, business people, famous actors and anybody who could advance or harm his career, and used these files for blackmail.

Try to figure out Mr. Hoover's imprinted and conditioned selves, according to the above analysis.

2. Try the same on Jesus Christ.

3. Try Thomas Jefferson.

4. Let each member of the study group pick some subject, or victim — not part of the group, but someone the member sees daily. Let the member study that person carefully and analyze which selves appear most often, how frequently the selves shift, and which self (if any) appears dominant most of the time.

5. This exercize will seem the hardest in the book, but try it anyway. Observe yourself for a week, and try to see which selves appear most often, if one self appears dominant, etc.

NINETEEN

Multiple Universes

The quantum theory of observer-created universes has implications far weirder than we have discussed thus far.

Some physicists do not agree with the Copenhagen Interpretation. They believe that we *can* make statements about a "deep reality". Unfortunately, the statements they make usually sound like either science-fiction or Oriental mysticism.

We will consider the "science fiction" flavors of quantum theory first. These began, in 1935, when Nobel laureate Erwin Schroedinger posed the problem we have already mentioned several times — the case of the cat which occupies the categories of life and death simultaneously.

Since quantum "laws" do not have the absolute nature of Newtonian (or Aristotelian) laws, all quantum theory must use probabilities. As we said earlier, the Aristotelian "yes" or 0 per cent and "no" or 100 per cent represent the certitude Occidentals have traditionally sought. Quantum experiments refuse to yield such certitude, and we find ourselves always with some probability between 0% and 100% — maybe 24%, maybe 51%, maybe 75%...etc.

In many cases we actually find a 50% probability, just as if we had flipped a coin and the chance of its landing heads = the chance of its landing tails = 50%. Schroedinger considers the case of a quantum decay process in which at any point in time, t, the chance of one possible outcome = the chance of another possible outcome = 50%. For convenience, let us make t, our time = 10 minutes. We can now say that after ten minutes the chance of outcome A and outcome B both equal 50%, but we cannot say which outcome will occur until the 10 minutes pass and we make a measurement.

Now, Schroedinger says, imagine a poison gas pellet which will explode in the case of outcome A but will not explode in the case of outcome B. Obviously, at the end of 10 minutes, the chance of an exploded pellet = the chance of an unexploded pellet = 50%.

Now put the pellet in a box with a cat, and lock the door. *Until the moment you open the door to see what happened*, the chance of the exploded pellet still = the chance of an unexploded pellet = 50%. Therefore, the chance of a live cat = the chance of a dead cat = 50%.

In Aristotelian language the cat "is" both dead and alive until we open the door.

Reformulating in operational language, as we have done, saves us from that absurdity, but does not entirely solve our problem here. The model that contains a dead cat has the probability of the model that contains a live cat and both still equal 50%. We seem to have escaped from the more bizarre metaphysical interpretations of the Schroedinger's cat problem, but we still have a mystery on our hands. Classical physics could predict *exact results* even *before* we looked, but quantum physics can only predict *probabilities until* we look.

Good grounds for sticking to the Copenhagen Interpretation, you might say. I tend to agree.

But Einstein and others did not like the Copenhagen view and kept insisting that eventually we would find a way to make statements about "reality", even in the quantum realm. Schroedinger's Cat made severe problems for them.

In 1952, Hugh Everett of Princeton, in collaboration with Wheeler and Graham, proposed a theory which attempts to describe "reality" and includes an answer to the Cat mystery.

In technical language, the probability wave that describes the possible outcomes of a quantum process has a mathematical name — the state vector. In the Cat problem, the state vector, as defined, can "collapse" two ways, one yielding a dead cat and the other yielding a live cat.

Von Neumann would say that, until we open the door, the state vector has the three values of dead cat, live cat and *maybe*. This means the state vector can collapse two ways and we remain in "maybe" until we actually see how it collapses in a given case.

Everett, Wheeler and Graham offer a different model, now abbreviated the EWG model, after their initials.

In this model, the state vector never "collapses." Each possible outcome does occur, in different *eigenstates* (roughly, probability manifolds). Since these eigenstates must exist somewhere, and cannot co-exist in the same space-time, they exist in different universes.

Thus, in super-space — a concept invented by Wheeler to solve quite different problems (mathematical formulations of gravity in Einstein's universe) — our universe does not exist alone. Other universes also exist — an unknown number of them — in the same super-space that contains Einstein's four-dimensional universe. In one universe, the eigenstate at the end of the poison pellet case contains a dead cat. In another universe, the eigenstate contains a live cat.

And this happens every time a 50% probability occurs. The state vector "splits" into two vectors in two universes.

Thus, this theory literally means that somewhere in super-space, a universe exists with an Earth just like this one, except that Adolph Hitler, over there, never went into politics and remained a painter, but Van Gogh, after his brain had collapsed from paresis, did enter politics and emerged as a Great Dictator.

If you depart from physics for a moment and get philosophical, or morbid, about this, you might ask why we had the bad luck to land in this universe instead of that one. The answer, in the EWG model, says that "we" exist in that other universe, too — or something like Xerox copies of us exist over there.

Well, I warned you this would sound like science-fiction.

Somewhere in super-space exists a universe in which Earth revolves around the sun just as it does here, but life never appeared on that Earth — and nobody like Everett, Wheeler and Graham ever evolved to suggest that other Earths existed, including one where our Everett, Wheeler and Graham dreamed up this idea.

Somewhere in super-space exists a universe in which a Xerox copy of me sits writing this paragraph and offers as an example, "Somewhere in super-space exists a universe where Beethoven died comparatively young, so they have the First to Fourteenth symphonies, but never heard the glorious Fifteenth."

Somewhere in super-space exists a universe in which another Xerox copy of me sits writing, "Somewhere in super-space exists a universe where Beethoven died very young, so they have the First to Fourth symphonies but not the glorious Fifth."

And so on...but not to infinity. Nobody has calculated the precise number of parallel universes that should exist, according to this model, but since all possible universes have to emerge from the same Big Bang (within the terms of this model) the number seems very large, but not infinitely large. Dr. Bryce de Witt in *Physics Today*, 1970, estimated it as larger than 10^{100} but did not attempt to guess how much larger.

Still, that's large enough for any science-fiction plot you care to imagine.

In one universe, presumably, I felt compelled to write a book much like this one, but due to different imprints and learning experiences, I rejected the Copenhagen view and the whole book consisted of an argument that the multiple-universe theory makes more sense than any other interpretation of quantum mechanics.

In one universe, whatever you think of this theory here, you think the opposite over there.

The major argument for the EWG model lies in its alleged *economy*. You may find this startling. Occam's razor, as every school child once knew — back in the reactionary days when schoolchildren had to know something — holds that scientifically, we must always choose the most economical model, the one which includes the least assumptions or presumptions. Now, on the face of it, a model which says that more than 10^{100} Xerox copies of you will read an equal number of variations on this text does not seem very economical. But the EWG advocates insist that all rival interpretations lead to even less "economical" conclusions.

The Copenhagen interpretation, for instance, seems much more economical than EWG — as I have presented it. However, all too often physicists have presented it, not in E-Prime but in standard English, including the "is of identity". *When stated with the "is of identity" the Copenhagen view always seems to say that we literally create the physical universe by observing it* — a position previously espoused only by Bishop Berkeley, and easily caricatured as solipsism. (As noted, one physicist has even written, "There is no reality.")

Thus, the EWG crowd claims that Copenhagenism violates Occam's economy by postulating *a universe magically created by human thought.* Because of the "is of identity" some Copenhagenists have actually gone that far. This led to Einstein's famous sarcasm that every time a mouse looks at the universe the universe must change; and Dr. Fred Allan Wolf has solemnly replied that the cells in the mouse's brain number so few that all the changes caused by all mouse observations total very, very little more than 0% and hence we can ignore them.

I think Copenhagenism, as expressed in this book, without the "is of identity" evades the above criticism. (We will shortly ponder whether another alternative, hidden variable theories, can similarly evade the EWG criticism when restated without the "is of identity".)

However, I will grant that *the EWG model does accept the basic wave equations of quantum mechanics at face value* whereas the EWG and hidden variable models add philosophical interpretations on top of the equations. In that sense the EWG model may qualify as more economical.

You see, the position of this book does not embrace what I call Fundamentalist Copenhagenism — the view that the Copenhagen model says the last word forever. Rather, I consider my position Liberal Copenhagenism. I do not believe any model equals the universe, or universes, but I think alternative models will continue to proliferate, because the data of modern science has grown so complex that many models will cover it.

Some call this Liberal Copenhagenism "model agnosticism." Dr. Marcello Truzzi calls it "zeteticism."

The history of the EWG model indicates the extent of fundamental disagreement among physicists about these matters and thereby, I think, reinforces the Liberal Copenhagenism or "model agnosticism" to which most physicists have, by now, retreated — the zetetic attitude of this book. Dr. Wheeler, one of the inventors of EWG, later rejected it for its "excess metaphysical baggage" but more recently he has returned to it again. Dr. Bryce de Witt says he could not take EWG seriously at first, but has now become one of its leading advocates. The majority of physicists still regard it as mathematical surrealism but its popularity continues to grow among the younger generation. At least a dozen books in the last decade have either espoused the EWG model overtly, or treated it respectfully — as just as plausible as the dominant Copenhagen theory.

We now see that, just as current neuroscience denies one essential self or "soul" of the Aristotelian sort and detects a variety of selves in every brain, one branch of quantum theory also sees a variety of selves. In other words, both brain research and one flavor of quantum mechanics say many possible selves appear equally "real" — the neurologists find these selves in our brain chemistry and the EWG theorists find them in other universes, but in both cases *the "one essential self" has vanished as totally as in Buddhist theory.*

In neuroscience the predominant self of the moment appears no more real than the latent selves which might manifest as soon as I take a drink or a drug, or get frightened, or find myself in an unfamiliar country. In the EWG model, the self which I manifest in this universe appears no more real than numerous female selves I have in half of the possible universes, or the countless alternative male selves I have in other universes.

One cannot help being struck by the fact that according to Freudian, Jungian and Gestalt methods of dream interpretation, these alternative selves, some of them bringing alternative universes with them, manifest every night in our sleep. Some physicists describe the other universes and other selves as "virtual", but does that not also describe our dreams?

And does it not appear that virtual selves and virtual realities have infiltrated both psychology and physics because, as this book claims, *all sufficiently advanced analysis must eventually abandon Aristotelian certitude and accept models — reality tunnels — based on probabilities?*

Exercize

Let each member of the study group say aloud "I do this exercize because..." and then attempt to state "all" the reasons. For instance, you will do this exercize because you have entered this class. Why did you enter this class? How did you get interested in the topics discussed in this book? How did you find this class? How did you arrive in this particular city of all the cities on this planet?

Carry the analysis further. How did you happen to be born? That is, how did your parents come to meet and mate? How did they come to be born? Amid all the wars, earthquakes, famines and other disasters of human history how did those genetic strains which combined in you survive when so many other genetic strains disappeared?

How did this continent emerge from geological evolution? Can you estimate how many migrations, wars of conquest, economic upheavals, etc. led to the genetic strains of your father and mother coming together?

Attempt in at least a rough, vague way to account for the formation of the planet Earth and the appearance of life on Earth.

When each member has had a chance at this game consider the *improbability of all of you coming together on this night, of all the nights of your lives, to do this exercize.*

It will probably prove necessary to do this exercize at least three times before the full meaning sinks into the neurons.

TWENTY

Star Makers?

Dr. John Archibald Wheeler has a more radical view of the matter these days than he had back when he co-authored the EWG model.

But before discussing that, we need to look at "non-locality."

In 1965, Dr. John S. Bell published a paper which physicists refer to tersely as "Bell's Theorem." Since a great deal of nonsense has gotten printed about this — and I wrote some nonsense myself in an early book called *Cosmic Trigger I: Final Secret of the Illuminati* (Falcon Press 1987) — we will take this very slowly. Bell's Theorem asserts that:

If some sort of objective universe exists in some sense (i.e., if we do not accept the most solipsistic heresies uttered by careless proponents of Copenhagenism), and,

If the equations of quantum mechanics have a similarity of structure (isomorphism) to that universe, then,

Some sort of non-local correlation exists between any two particles that once came in contact.

The full weirdness of this will strike you when you remember that the classic type of non-local correlation previously claimed among humans consists of the "magical" idea that if a shaman gets his hands on a lock of your hair, anything he does to that hair will have an effect on you. Frazer called that idea "sympathetic magic" in *The Golden Bough* and characterized it as typical of "primitive" thinking. Has the most advanced science returned to the most "primitive" ideas?

Not quite. We will explain the subtleties of "non-local correlation" in a moment. First, let us note that the idea of non-local correlation seems so unholy or unthinkable to some physicists (who recognize its haunting resemblance to shamanic magic) that they have decided to evade the consequences of Bell's math by challenging the first step above and retreating to an unashamed solipsism. This path has thus far appeared overtly (as far as I know) only in two articles by Dr. N. David Mermin of Columbia University ("Quantum Mysteries for Everyone," *Journal of Philosophy* Vol. 78, 1981, and "Is the Moon There When Nobody Looks?" *Physics Today*, April 1985). Dr. Mermin claims the moon does disappear when nobody looks.

Yes. I do not exaggerate. Dr. Mermin writes, "The moon is demonstrably not there when nobody is looking." Ah, I think, if only the man had some knowledge of E-Prime...

Please remember that Dr. Mermin's position differs from my claim, which holds that the moon does not appear in *our observed universe* until somebody looks, but I do not assert we can make meaningful assertions about either *existence or non-existence* in "the real universe" and can only make meaningful utterances after somebody looks at the observed universe.

Nobody has cared (yet) to challenge the middle term in Bell's argument. The equations of quantum mechanics have higher isomorphism with the observed universe than anything else in science. We know this because these equations appear in the theory underlying about 90% of the technology we use every day (estimate of John Gribbin). They appear in TV, in atomic energy, in computers, in molecular biology, in genetic engineering and "all over the shop." If these equations had a major defect, it would have come to light by now. (Minor defects probably exist, as in all human endeavor, but a major defect would mean that things would blow up all around us every day.) No: the quantum equations have, probably, the highest level of pragmatic (experimental, practical, daily use) confirmation of any branch of science.

So, assuming a universe that humans can observe, and an isomorphism between that universe and quantum mathematics, Bell's conclusion seems mathematically inescapable. It has also survived seven experimental tests, with increasingly sophisticated instruments, and seems vindicated to all but those who, like Dr. Mermin, find solipsism less "irrational" than non-locality.

So, what do we mean by "non-locality" in Bell's sense? Can we differentiate it from shamanic magic? Yes, we can. It will then appear, not at all "as weird as magic." It will appear far weirder.

All pre-quantum models of the universe, including Einstein's Relativity, have assumed that *all correlations involve connections*. In other words, they assume that if whenever A goes **ping!**, B then goes **pong!**, the explanation must lie in some connection between A and B. If the ping-pong response continued, over and over, with no connection between A and B, that would seem spooky indeed to classical physics (and to common sense).

In Newtonian physics, the connection between ping and pong appears mechanical and deterministic (A pushes and B gets pushed, etc.); in thermodynamics, the connection appears mechanical and statistical (when enough A's bounce around enough, they will hit enough B's to get the B's bouncing, too); in electromagnetism, the connection appears as the intersection or interaction of fields; in Relativity, the connection appears as a result of the curvature of space (which we call "gravity"); but the correlation, in any case, involves *some sort of connection*.

In a simple model, all pre-quantum physics assumed a kind of billiard-table universe. If a ball moves, the cause lies in mechanics (it got hit by another ball) or fields (an electromagnetic field pulled the ball in one direction rather than another) or geometry (the table curves a certain way) but the ball does not move without cause.

In quantum mechanics, since the 1920s, non-local effects — correlations without connections — have seemed to many physicists the only explanation of some of the behavior of sub-atomic systems. (Bohr used the word "non-local" as early as 1928.) Bell merely proved mathematically that these non-local effects indeed must exist if quantum math meshes with the observed universe.

In these non-local effects, when we say no connection exists to explain the correlation, we mean, more bluntly, that no "cause" exists — in any sense that we have ever understood "cause".

Imagine a billiard table without players. Nobody hits any balls. No earthquake shakes the room. No magnet exists, hidden under the table. Yet suddenly Ball A at one end of the table turns clockwise and Ball B at the other end of the table turns counterclockwise.

If you told that to the Incredible Randi, he would insist some fraud or hoax existed. Yet such non-local correlations appear mathematically necessary to quantum mechanics and experiments have measured them repeatedly.

The billiard table model only suggests one aspect of non-local reality. Another model, from a lecture by Dr. Bell himself — as reported to me by Dr. Herbert — goes as follows:

Imagine two men, in Dublin and Honolulu. Imagine that we have observed them carefully for some time and have deduced some "laws" of their behavior. One law states that, whenever the man in Dublin wears red socks, the man in Honolulu will wear green socks. We then experimentally meddle with the system — we "cause" ("or leastways bring it about," as Joyce says) that the man in Dublin takes off his red socks, and dons green socks. We immediately check our monitors in Honolulu. We find that the man there has *instantaneously* taken off his green socks and donned red socks! (I disapprove of exclamation marks in expository prose, but this case seems to deserve at least one. Perhaps it deserves three or more…)

"Instantaneously" means, among other things, that, we know for sure that no signal from Dublin could have reached Honolulu to create a connection between the events. *Signals travel at the speed of light (or less) and cannot cause an instantaneous response.* So the result in Honolulu does not even qualify as a "response", strictly, and would quickly get classified as a coincidence — except that, if these men continued to act like Bell particles, the same correlation would occur every time we got either men to change his socks.

(The highly technical atomic experiments showing this kind of behavior appear in Gribbin's *In Search of Schroedinger's Cat* and Herbert's *Quantum Reality*.)

Well, now, you can see how this differs from primitive "sympathetic magic". Magic involves some "occult" theory of causality, but this correlation without connection does not fit *any* theory of causality. Magic also travels one way, in theory, but this does not seem to involve travel at all — unless you want to try to imagine two-way instantaneous "travel". In short, magic does not have quite the weirdness of non-local correlation.

Dr. Jack Sarfatti, incidentally, calls the non-local correlation "information without transportation." You might try that if you find my term "correlation without connection" a bit opaque.

Jungian synchronicity, of course — accepted not just by Jungians but a lot of other psychologists — also involves this kind of non-local and non-causal correlation. Indeed, Jung specified that synchronicity could not fit into any purely causal, billiard-ball theory of the universe. Most scientists outside psychology felt, before experimental verifications of Bell's Theorem, that only psychologists could talk such nonsense... But now the matter seems to need re-examination.

Let us now, finally, confront the implications of Bell's Theorem for the multiple universe model.

In the last decade, physicists have spent a lot of time debating something called the Anthropic Principle which says, briefly, that we live in a universe that looks suspiciously as if it would *necessarily* produce human beings eventually. In less careful language, it looks as if "designed" for humans.

Now, this reverses the last 300 years of science. "Design" in general and especially anthropic design played a large role in Aristotelian and theological thought once, but science decided it could do very well without any "design" postulates. Nonetheless, the anthropic principle now seems stronger that it ever did in the theological era.

You see, several cosmologists have noted a group of rather singular facts about our universe, all of which reduce to the simple proposition: *if we change any of the constants of physics even very slightly (as little as 0.01 per cent in many cases) we find the result would produce a universe in which humans could not have evolved.*

In other words, of all possible universes that might have emerged from the Big Bang, most of them would either collapse quickly, or evolve briefly into various gaseous formations, or evolve into galaxies of stars without any planets, or develop in one way or another that would not allow the possibility of human life.

Prof. Paul Davies examines these demonstrations at length, with mathematical rigor, in his *The Accidental Universe*. He arrives at the conclusion that we must either accept the EWG model of many, many universes, most of them without humans, or, *if we insist on the "common sense" one universe,* we must accept that some Anthropic Principle has worked to "design" or evolve that universe in a direction that made it possible for us to exist.

In simple language — we have two choices: many universes, or one universe with something suspiciously like a Designer. No matter how hard its proponents work to make the latter choice sound abstract and mathematical, it still sounds like "God" to most readers.

Of course, the Designer (as conventionally conceived) does not receive a cordial reception in scientific circles. He belongs to the theologians, scientists think, and He definitely does not belong in a scientific account of the universe.

In current debate, the Anthropic Principle has broken down into a Strong Anthropic Principle and a Weak Anthropic Principle; the former yields results more compatible with the Designer hypothesis. Even the Weak Anthropic Principle, however, opens a door through which the Designer may creep back into science.

Now Dr. Wheeler re-enters the story.

(In this discussion, I largely follow a popularization of Dr. Wheeler's current theorizing: "Turning Einstein Upside Down," by John Glidedman, *Science Digest,* October 1984.)

Bell's Theorem shows that, if quantum theory corresponds to something like a physical universe — if quantum theory does not collapse into solipsism, as critics of Copenhagenism always expect it will — non-local correlations must exist in the universe as well as in our math. But these non-local correlations need not be *correlations in space,* as I have presented them thus far (for the sake of clarity and simplicity). Bell's Theorem indicates that non-local *correlations in time* must also appear in a quantum universe. (I left that out until now because it really boggles the mind, if the reader does not get led to it slowly, a step at a time.)

Non-local space-correlations (or space-like correlations, as strict Relativists would say) merely abolish our concept of linear causality, or transcend it, or ignore it. Non-local time-correlations, or time-like correlations precisely turn causality on its head.

Thus: two photons enter the same measuring instrument. This creates the contact that becomes, in Bell's math, a non-local correlation. But one of the photons came to the instrument from a candle across the room, and the other came from a star 1,000,000 light-years away. But the non-local correlation does not change (its rate of change = 0, in Bell's equations) in either space-like or time-like separations.

For the two photons to fit this requirement, the one that left the star 1,000,000 years ago must have had its properties set in place 1,000,000 years ago, which seems absurd, even for the Quantum Wonderland. (This implies the photon "knew" we would measure it 1,000,0000 years later, so it dressed appropriately before leaving the star and beginning its long journey.)

So, then, alternatively, the photon doesn't leave the star 1,000,000 years ago "until" in a sense the result of our measurement today travels non-locally in time "back" to the star and "adjust" the photon to correlate with the other photon from the candle.

What did I just say?

Yes, we now have a backward-in-time causality, not necessarily as the literal truth in some Aristotelian sense, but as *the only kind of model that makes sense in terms of the data we now have.*

In Wheeler's words, "...we are wrong to think of the past having a definite existence 'out there'."

Copenhagenistically and pragmatically, any model of the past serves us, or fails to serve us, in dealing with our problems now. The traditional model of the past — having a definite existence "out there" — does not serve us, if we want to understand the non-local correlation. Thus, we need a model in which the present can influence the past.

Without the "is of identity", this poses no problems, for me.

With the "is of identity", we arrive at endless paradoxes and a model that only a mental patient can take seriously. We arrive at, in fact, a universe which changes *as a block-like entity — all of it, past, present and future — every time we make a measurement.*

And, I think, if Einstein's mouse accidentally triggers our measuring instruments, then the mouse can, after all, change the whole universe.

Dr. Herbert sensibly asks how taking a measurement can have this "magic" power. I don't think it can. I think we need a model with backward causality, at this point, until we find a better model, because otherwise we contradict the facts of quantum experiments. But I do not think the model "is" the universe. When the model gets this peculiar, we need to build a better model.

Meanwhile, until the better model (or "new paradigm") arrives, *within the current Wheeler model,* the many, many universes of the old EWG model still exist in super-space somewhere, but we have "selected" this Anthropic universe by the kind of experiments we have conducted.

If you accept the Wheeler model as the final truth, then...

The Designer, at long last, appears revealed; the door of the Law has fallen open and we enter.

Our experiments here and now, Wheeler says, travel non-locally in time-space, as Bell indicates. Along the way they intersect the Big Bang, along with everything in general. The Big Bang thus gradually gets "fine-tuned" so to speak and the universe around us becomes Anthropic — a universe in which humans can and must exist. *We did it to ourselves.*

In short, we don't need to postulate a supernatural Designer. Our experiments create the universe observed by our experiments — which when interpreted always yield an Anthropic universe, rather than any of the millions or billions of possible non-Anthropic universes — because we designed the experiments.

Okay?

Of course, this still does not mean the same as the New Age slogan "We create our own reality." Wheeler emphatically does not think *thought or mind or consciousness* have anything to do with this circular-causal chain. Only nuclear experiments influence the particles which non-locally tune the Big Bang to produce our selves and our universe.

Nonetheless, it seems decidedly odd that the "Designer" also appears as you and me and the guy leaning on the lamp-post in the account of quantum mechanics written by Eddington nearly 60 years ago, and Eddington did not follow Wheeler's path of non-local correlations and backward-in-time causality. Eddington merely followed the Copenhagen Interpretation back into its origins in pragmatism and existentialism, as I have, and arrived at the conclusion which he expressed thus *(Philosophy of Physical Science,* page 148):

> We have found a strange foot-print on the shores of the unknown. We have devised profound theories, one after another, to account for its origin. At last, we have succeeded in reconstructing the creature that made the foot-print. And, lo! It is our own.

The Sufi equivalent, a thousand years earlier, goes like this: the marvelous Mullah Nasrudin, while out riding in the desert, saw a band of men on horses in the distance. Knowing that bandits frequented that area, Nasrudin galloped off in the opposite direction, as fast as his donkey could travel.

The men on horses, however, recognized the divine Mullah. "Now why would the wisest man in Islam rush off like that?" they asked one another. So they decided to follow him, thinking he would lead them to something marvelous.

Looking back, Nasrudin saw that the "bandits" had started to chase him. He spurred his donkey to gallop faster. The men followed faster, determined not to miss out on the mysterious doings of the great Nasrudin. The chase continued, with everybody rushing faster all the time, until Nasrudin saw a graveyard. Quickly, he dismounted and hid behind a gravestone.

The men rode up and, sitting on their horses, looked over the gravestone at Nasrudin. A thoughtful pause occurred, while everybody pondered hard, especially Nasrudin who now recognized the men as old friends. "Why are you hiding behind a gravestone?" one of them finally asked.

"It's more complicated than you realize," Nasrudin said. "I'm here because of you and you're here because of me."

Exercizes

1. Let the class discuss the Zen riddle, "Who is the One more wonderful than all the Buddhas and sages?"

2. According to a story in *News of the Weird, op. cit.* six men in the Philippines once got into an argument about "which came first, the chicken or the egg." Tempers flared, guns emerged, and four of the six got shot dead. See if the class can discuss the Wheeler theory, pro and con, without equally drastic results.

3. Apply, with your own *ingenium*, the Wheeler model to an ordinary quarrel between humans.

4. Take the top off the tank behind the toilet, pull the handle and watch how the water level returns to its previous height after flushing. This shows the simplest possible circular-causal mechanism in an ordinary home. Apply circular-causal analysis to:

    A. Race relations in the U.S. and the Union of South Africa.

    B. The cold war.

    C. The average divorce.

    D. Self-fulfilling prophecies in corporation/union relations.

TWENTY-ONE

Wigner's Friend, or Whodunit?

Another Nobel laureate, Dr. Eugene Wigner, added a more complex twist to the Schroedinger's Cat problem, and conclusions emerged similar to Wheeler's observer-created Anthropic universe, but also startlingly different.

Please remember that we deal always with probabilities, not certitudes, and you will not get too flustered as we proceed to the next twist in Quantum Psychology's kinky yellow brick road.

In the original cat problem, we had a physicist in a lab, a box, a cat inside the box, a poison gas pellet inside the box, and some radioactive decay process that would sooner or later trigger the explosion of the pellet and the death of the cat. We found that, without opening the box, the equations that describe quantum decay yield a solution in which the statements "the cat has died" and "the cat still remains alive" remain equally "true" or equally "false", or at least remain 50% probabilities. Von Neumann's logic would have us say both statements remain in the "maybe" state, like a coin in mid-air.

When we open the box, we find a live cat, or a dead cat, and no more *maybe*, like a coin that has landed heads or tails. It seems then that *we collapsed the state vector by opening the box.*

Very well, but now let us look at it from the perspective of another physicist, outside the laboratory. Wigner called this second Observer a friend of the physicist in the lab, and thus this new problem has the title, The Wigner's Friend Paradox.

After ten minutes, as in our original example, the physicist in the lab, Ernest, opens the box and finds a live cat. (I like happy endings.)

For Ernest, then, the state vector has "collapsed". The probabilities no longer register as 50% dead and 50% alive, but as 0% dead and 100% alive.

The friend in the hall, Eugene, however, has not heard the news yet. From his perspective, Ernest in the lab remains, like the whole experimental system, in a "maybe" state. Very concretely, Ernest consists of molecules, which consist of atoms, which consist of "particles" and/or "waves", which follow quantum laws, and Ernest remains with an uncollapsed state vector...until he opens the lab door, sticks his head out and announces, "Tabby hasn't died yet." For Eugene, then, *hearing the news collapsed the state vector.*

Of course, we all consist of molecules, which consist of atoms, which consist of "particles" and/or "waves" and we all remain in various "maybe" states until we make a choice in the existential sense.

Between choices, we evidently return to the "maybe" state until we make another choice. *"Existence precedes essence,"* remember?

So, from the point of view of Eugene in the hallway, we all contain quantum uncertainty. The quantum uncertainty only "collapses" into a definite "he done it" or "he didn't do it" when Eugene observes us.

Now, across the ocean another physicist waits impatiently for the result of this experiment in felixicide. Let us call her Elizabeth.

From Elizabeth's point of view, the state vector does not collapse when Ernest tells Eugene, "Got a live cat in here, after all." The state vector in Elizabeth's universe only collapses when Eugene rushes to a phone, a fax, a computer net or whatever and transmits the signal, "Live cat this time." For Elizabeth, *the state vector collapsed when the signal arrived.* The signal, then, collapsed the state vector, in Elizabeth's universe.

A fourth physicist, Robin, waits anxiously to hear what electronic message Elizabeth has received...and in Robin's world, the state vector has not collapsed yet...

And so on...for any number of Observers.

We seem to have arrived back at Von Neumann's Catastrophe of the Infinite Regress, in a different form.

Some will attempt to evade the obvious implications here by saying the state vector only exists as a mathematical formula in human heads — and only in some human heads (those belonging to physicists, in fact). In that case, the Wigner's Friend problem does not have the radical import of Einstein's discovery of the relativity of instrument readings. In the Wigner case, the relativity (of when the state vector collapses) only exists in our conceptualizing, whereas Einstein's relativity exists in meter readings.

This objection overlooks the fundamental discovery of quantum mechanics, which I have stated in dozens of ways ever since the beginning of this book, but which remains so "alien" to our Aristotelian culture that we continually forget it even after we think we have learned. That discovery, to say it again and yet another way, consists in the facts that:

(1) We cannot make meaningful statements about some assumed "real universe", or some "deep reality" underlying "this universe," or some "true reality", etc. *apart from ourselves and our nervous systems and other instruments.*

Any statements we do make about such a "deep" reality separate from us can never become subject to proof, or to disproof, and that makes them "meaningless" (or "noise").

(2) Any meaningful scientific or existential or phenomenological statement reports on how our nervous systems or other instruments have recorded some event or events in space-time.

The reader has read several variations on this, and performed exercizes designed (I have hoped) to make this experientially clear, and yet Wigner's argument probably still sounds a bit "queer" to some of you out there. Well, *it concerns only probabilities* as I said again at the beginning of this chapter, and (1) it does not attempt to describe a "deep" reality separate from us and (2) it does describe the kind of "reality" we can experience with our nervous systems and other instruments, so the Wigner argument qualifies as meaningful scientific speech.

Let us try it in reverse. The "common sense" verdict would say, "Well, the damned cat is either dead or alive, even if nobody ever opens the box."

Since, by its own terms, this can never become subject to test, it has no meaning. As soon as a test does occur, and somebody peeks into the box, we have left "common sense" and/or Aristotelian "reality" and entered operational non-Aristotelian "reality". In short, once a test occurs we enter the area of science, of existentialism, and of meaningful speech. Without a test, we remain in the area of noise — "sound and fury signifying nothing," as the Bard said.

"The cat is alive or dead even if nobody looks" has an uncanny resemblance, if you think about it, to that other famous "isness" statement, "The bread is now the body of Jesus Christ, even if every instrument still registers it as bread." Such non-instrumental, non-existential "truths" may make good surrealist paintings or poems — they may provoke creativity and imagination, etc. — but they do not contain information or meaning in any phenomenological context.

But *"we"* of course remains undefined above.

If we define "we" as the folks in the laboratory, then meaningful speech begins when the box opens. If we define "we" as the folks in the hall, rubbing shoulders with Wigner's Friend, meaningful speech begins when Ernest opens the lab door and says "Live cat again." If we define "we" as the physicists across the ocean, meaningful speech begins when the electronic signal arrives...

I know, I know. It all sounds very weird.

That's why Einstein had to remind us, "Common sense tells us the earth is flat."

Please note, that, *even if "The cat is dead or alive even if nobody looks" may fit the meaningless category, but "The cat is dead" and "The cat is alive" do not fit that category.*

They fit the category of the indeterminate. Remember the distinction between the indeterminate and the meaningless?

Thus, "Somebody put a time bomb under the table" does not qualify as a meaningless statement, even if nobody has looked yet. The odds seem very high that somebody will look, if you speak this aloud. In fact, probably everybody will look...

The statement remains "indeterminate" in the time between hearing it and actually checking under the table carefully. Then it becomes either "true" or "false".

Got it?

Non-Aristotelian logic deals with existential/operational probabilities. Aristotelian logic deals with certainties, and in the lack of certainties throughout most of life, Aristotelian logic subliminally programs us to invent fictitious certainties.

That rush for fictitious certainties explains most of the Ideologies and damned near all the Religions on the planet, I think.

Exercizes

1. Classify the following propositions as true, false, meaningless or merely currently indeterminate.
  A. The U.S. Air Force has several dead extra-terrestrials hidden in a hanger at Edwards Air Force Base.
  B. This exercize contains thirteen propositions.
  C. All propositions in this exercize are false.
  D. No good cop ever takes a bribe.
  E. The function of public education consists of killing curiosity, encouraging docility, and preparing mindless drones to work for corporations.
  F. Gorbachev has an advantage over everybody else in the Politburo because he remains sober when the rest of them have all gotten drunk.
  G. Proposition B is false.

H. Proposition G is false.
I. God loves everybody, even serial killers, rapists and C.I.A. agents.
J. All propositions are true in some sense, false in some sense, meaningless in some sense, true and false in some sense, true and meaningless in some sense, false and meaningless in some sense, and true and false and meaningless in some sense.

2. Try living for one day with this (possibly) self-fulfilling prophecy: "I am dumb and unattractive and nobody likes me."

3. Try living for one day with this program: "I am brilliant and attractive and everybody likes me."

4. Decide which of the two above exercizes you liked best, and try living with that program for a full month.

Observe all old programs that re-assert themselves and interfere with this exercize.

PART FIVE

The Non-Local Self

If quantum mechanics hasn't profoundly shocked you,
you haven't understood it yet.
— Niels Bohr

TWENTY-TWO

Hidden Variables & The Invisible World

As I mentioned earlier, Einstein did not like the Copenhagen Interpretation. His debate with Bohr about this issue continued for over twenty years and filled the pages of many learned journals; the majority of physicists, when the debate had run its course, decided the Bohr had won. Nonetheless, some of Einstein's arguments continued to haunt the physics community and a small minority kept wondering if the Father of Relativity might not have scored a few telling points along the way.

Einstein's favorite line of criticism revolved around his claim that quantum mechanics, as known then (and as still known) may not constitute a "complete" theory of the subatomic realm. In ordinary language, this means that the Uncertainty and Indeterminacy of quantum equations — however useful these equations prove every day in technology — contains a possible hole through which an entirely new Quantum Theory may someday march.

In simple terms, the fact that we cannot remove the Uncertainty and Indeterminacy today does not necessarily register a fact about the limits of scientific method (as Bohr believed) or about the limits of scientific method and the human nervous system (as I have argued). Uncertainty and Indeterminacy may simply register the "incompleteness" of quantum mechanics.

Eventually, the Einstein argument evolved into the Hidden Variable hypothesis. Suppose we eventually discover variables, currently unknown, and suppose these variables explain the collapse of the state vector. If that happens, then the Copenhagen Interpretation will become obsolete — along with von Neumann's three-valued logic, the multiple worlds model, and the monstrous progeny of Schroedinger's cat and Wigner's friend.

A realm of hidden variables — an invisible, sub-quantal "world" — *if we can ever demonstrate it in a laboratory,* would then explain how the state vector "collapses" from a probability, before measurement, to a certainty, after measurement. We can then say the Hidden Variables did it, and we don't have to say the act of measurement did it — or the *report* of the act of measurement did it, as in the Wigner argument. "Common sense" and maybe even Aristotelian logic can arise again from the graves to which physics consigned them in the 1920s.

Unfortunately, two major objections exist to the Hidden Variable model.

In the first place, Hidden Variable theories "sound wrong" and even smell wrong to modern scientists. They suggest Aristotelian "essences" and even Platonic "deep realities" and other metaphysical entities, or spooks. They even remind some scholars of the alleged "hidden essence of Jesus" which Catholics claim lies buried within something that appears only a piece of bread to our senses and instruments. In short, they have a distinctly medieval stench about them.

More technically, this objection states that Hidden Variables still remain indeterminate and sound almost as if they might remain indeterminate forever and thus deserve the curse of "meaninglessness" or "noise". After all, no matter how many experiments fail to find the Hidden Variables, the die-hard proponents of these spooks can still claim, "We just haven't looked in the right place yet." That path leads to endless philosophical debate, not to scientific operations.

The second objection to Hidden Variable theories seems even stronger. Scientists have used quantum theory for 90 years in some form or other, for 70 years in its (allegedly) complete form, and have not found any evidence for Hidden Variables at all, at all.

These arguments led physicists, or the overwhelming majority of them, to consign Hidden Variables to the dustbin, alongside Catholic "essences", the phlogiston and luminiferous ether theories and "Natural Law" (in the moral or political sense). We can do quite well without such spooks, the scientific community agreed.

Or so it seemed until Dr. David Bohm suggested a new test for Bell's Theorem and Dr. Aspect of Orsay performed the test several times.

It now appears that we may have a more "complete" quantum theory at hand, one that includes Hidden Variables. However, at this point, that does not mean we have found "deep reality" and can junk the Copenhagen Interpretation. It simply means that we have another new model — which implies, for most physicists, another argument for "model agnosticism" or zeteticism.

Let us look at this Hidden Variable model:

Everybody agrees — well almost everybody: a few heretics dissent from every verdict in quantum mechanics — but *almost* everybody agrees that Aspect's experiments clearly demonstrated the non-local correlation.

Some claim Aspect also demonstrated a kind of Hidden Variable. Others vigorously deny this. John Gribbin, physics editor of *New Scientist*, claims that Aspect's experiments not only fail to support Hidden Variable theories but clearly refute them once and for all.

Obviously, we here enter an area where the best-informed physicists have trouble understanding each other, or what they mean by the terms of their own debates.

The problem seems to lie in different concepts of what we mean by a Hidden Variable. Dr. Bohm, the man who suggested the design of the Aspect experiments, means something that Einstein and other proto-Hidden Variable theorists had not conceived. From Bohm's point of view, the Aspect experiments weaken the case for *local hidden variables*, but they tend to support the concept of *non-local hidden variables*.

And what, by all the pot-bellied Buddhas in Burma, do we mean by a non-local hidden variable?

Well, as explained in the last section, non-local correlations transcend causality and also subvert our traditional notions of "space" and "time". If two "particles" — or "events", or Whatnots — have a non-local correlation, in modern quantum theory, this means that they will remain correlated even when no signal, no field, no mechanical push-or-pull, no energy, no "cause" of any sort can travel from one to the other.

In the Aspect experiments, for instance, photons at two ends of an experimental apparatus retained their Bell correlation whenever Aspect took a measurement — and yet the measurements could only occur after the photons passed special switches that allowed measurement only in the last 10 nanoseconds of the experiment. Light would have taken 20 nanoseconds to travel from one photon to the other, and — as you probably have heard — no energy known to physics can travel faster than light. In other words, no physical energy could carry a signal from photon A to photon B, or from photon B to photon A, and make a *connection* that would allow us a *causal* explanation.

You can see that no local hidden variables can account for this. Hence, one form of the hidden variable theory definitely cannot be invoked to explain this "correlation without connection." That hardly "refutes" local hidden variable theories, but does show, once again, that quantum experiments have not yet revealed any "incompleteness" that we can fill in by positing a local hidden variable.

However, a *non-local* hidden variable might explain Aspect's results and several other experiments, performed since Bell published his Theorem, which all indicate that non-local correlations do appear in the laboratory (experimentally) as well as in the equations (theoretically).

We still don't have a very clear idea of what a non-local hidden variable would look like, do we?

Dr. Bohm, who started thinking about non-locality as early as 1952, has over the years developed a mathematical model of non-local hidden variables, and — more marvelous yet — has even found a way to write in fairly normal English about what this math "means". (His

English involves turning a lot of static nouns into dynamic verbs, but I think I can get the same effect by merely continuing to avoid the static "is of identity.") You can read about Bohm's model (with his own peculiar English) in his book, *Wholeness and the Implicate Order* (London, Ark Paperbacks, 1983).

Briefly, Dr. Bohm posits an explicate or unfolded order (he uses both words) which makes up the 4-dimensional continuum known to post-Einstein science. This order, which we normally call the visible universe, he names explicate or unfolded because it occupies space-time — every part of it *has a location.* You can say you found the part here in space, not somewhere else, and now in time, not sometime else.

This explicate order corresponds roughly to the *hardware* of a computer, or to our brains. (Dr. Karl Pribram, the neurologist, has adopted Bohm's model to explain some mysteries of brain functioning.)

Dr. Bohm next posits an implicate or enfolded order (he uses both words) which both "permeates" and transcends the 4-dimensional explicate Einsteinian universe. This order he calls implicate or enfolded because it does not occupy specific space-time — *no part of it has a location.* You cannot find it only here in space, *but also somewhere and everywhere else;* you cannot localize it only here in time, but *anytime and everytime else.* (Only its explicate results have locality. It itself remains non-local.)

This implicate order corresponds to the *software* in our computers — and in our brains, according to Dr. Pribram.

Everything on the explicate-unfolded level has locality and appears causal (until one examines its smallest, quantum parts); everything on the implicate-enfolded level has non-locality and appears non-causal.

We have only managed to observe non-locality in the form of its results, i.e., non-local correlations on the explicate-unfolded (space-time) level — since Bell inspired us to look for them — because this unfolded level acts as the extension in space-time of the always non-local and non-spatio-temporal implicate order. In other words, a sub-quantal world rather like the "deep reality" banned by Niels Bohr exists, but we cannot observe or experience it; yet we cannot call it a "spook" or "meaningless concept" because we observe its effects as non-local correlations that make no sense at all unless we assume something like this implicate order.

However, the implicate order as a scientific model does not equal classic "deep reality" in the Aristotelian sense, because it has the role of one model among many. Bohm, its father, does not claim it ranks as "the only true model" or the "final" model or anything like that.

If the computer and brain metaphors have not made the implicate order clear enough for the reader, try another model: a performance of Beethoven's *Ninth* has all the characteristics of hardware or the explicate order. You can locate it very precisely in space-time — at 9 p.m. on Tuesday at the Old Opera House, say — and if you get these space-time coordinates confused you will miss the performance.

But Beethoven's *Ninth* also has an implicate, enfolded existence as software, which does not exactly correspond to Dr. Bohm's implicate order but approximates to it. If every printed copy of the symphony could have a date and locale affixed — "We found this one in Lenny Bernstein's summer home on Sunday November 23," etc. — some aspect of the *Ninth* would still remain non-local because we can't say, exactly, how many heads contain all or part of it.

In Ray Bradbury's science fiction novel *Fahrenheit 451*, a totalitarian state has burned all books, but this only destroys the local "hardware" of the books. A group of subversives have memorized all the classics, teach them continually to others, who teach others, etc. and the books remain partly non-local and inaccessible to the book-burners. (Some of us have similarly preserved some George Carlin routines, not out of any paranoid fear that our society will soon enter a similar totalitarian nightmare, but simply because we have played his videos so often that we have memorized large portions of them. Quite large parts of *Casablanca* have achieved a similar degree of non-locality.)

These musical-literary analogs intend to help the reader get a handle on non-locality. *But true non-locality in the Bohmian sense would continue even if all human beings died.* An approximation toward a thinkable model of this requires television as an exemplar. As cynics often note mournfully, the TV shows of the 1950s still travel through space-time and denizens of solar systems forty light-years away, systems we can't even see, might start receiving Ed Sullivan, Milton Berle and news of the MacCarthy Era any day now — and try to understand us on the basis of those signals...

Thus, Milton Berle has achieved something like non-locality.

Another analogy, without locality and non-locality, at least helps clarify implicate and explicate. I call you on the phone. The words emerge from my mouth as implicate or unfolded sound-waves. The transmitter in my phone converts them to implicate or enfolded electrical charges. The receiver on your phone picks up these enfolded charges and unfolds them, so they become explicate sound-waves again, and you "hear me talking."

Similarly, a friend in New York sends me a bit of software on a floppy disc. I put this enfolded message in my computer and it appears, unfolded, as a new computer game, on the console screen.

Now, the consequences of this implicate/explicate model seem even stranger than we suspect at first sight. For instance, just as Einstein's Relativity abolished the dichotomy of "space" and "time", and modern psychosomatic medicine tends to abolish the distinction between "mind" and "body", this Bohm model seems to undermine our traditional dualism of "consciousness" and "matter".

In a non-local implicate order, information cannot have a locality, but "permeates" and/or "transcends" all localities. And information that has no locality sounds a great deal like the Hindu divinity Brahma, the Chinese concept of Tao, Aldous Huxley's "Mind At Large," and "the

Buddha-Mind" of Mahayana Buddhism. Any one of those concepts must mean *information without location* (if we admit they mean anything at all).

"The Buddha-Mind is not 'God'," Buddhists continually explain, and Occidentals blink, unable to understand a religion without "God". But Brahma, in Vedic Hinduism, does not have any of the personality, locality, temperament (or gender) of Western "gods" and, like Buddha-Mind, seems to mean a kind of non-local implicate order, or information without location, if it means anything.

Bohm has avoided speculating about this parallel between his math and ancient Oriental mysticism, but others have not. Dr. Capra in *The Tao of Physics* uses a Bohmian non-local model of quantum theory as the "true" model (ignoring the physicists who prefer EWG or Copenhagenism) and then points out, quite correctly, that (if we accept this as "the only true quantum model") quantum theory says the same things Taoism has always said.

Indeed, Lao-Tse's famous paradox, "The largest is within the smallest" only begins to make sense to an Occidental after she or he has understood what non-local information means in modern physics.

Dr. Evan Harris Walker goes further. In a paper, "The Compleat Quantum Anthropologist" (American Anthropological Association, 1975) Dr. Walker — a physicist, not an anthropologist, by the way — develops a neo-Bohmian Hidden Variable model in which "consciousness" does not exist locally at all but only appears localized due to our errors of perception. In this model, our "minds" do not reside in our brains but non-locally permeate and/or transcend space-time entirely. Our brains, then, merely "tune in" this non-local consciousness (which now sounds even more like Huxley's "Mind At Large").

Dr. Walker develops a mathematical model of this non-local Self and uses the model to derive predictions about how often the alleged psychokinesis of parapsychologists can occur. His results correlate with the scores made by persons very successful in psychokinesis experiments. In other words, people rated good at "controlling" the fall of dice, because they score above chance, *score on the average only as far above chance at the non-local Hidden Variable model says they can.*

(For further details on the Walker model and its correlation with parapsychology, look up his paper, or consult my *New Inquisition, op. cit.*)

Exercize

Since you probably do not have the equipment to perform sub-atomic experiments, and since the "parapsychological" implications of Hidden Variable theories excite people more than the physics does, try the following experiment:

Secure a brain-machine that produces 4 hertz wave forms[1] in the brain. (You can obtain a catalog of brain machines from Inner Technologies, 51 Berryl Trail, Fairfax CA 94930.) Set the machine for 4 hertz for about 30 minutes. Then conduct some classic ESP and PK experiments in the class.

Try to avoid prejudice, pro or con PSI research. Try to just do the experiments and note the results.

I would consider it a great kindness if you would send me the results, whatever happens.

---

[1] Nobody knows why at present, but "paranormal" events seem much more likely at 4 hertz than at any other brainwave frequency.

TWENTY-THREE

Quantum Futurism

Earlier, we discussed the four basic systems that, in the majority of Terrans at this stage of evolution, make up the hardware and software out of which our multiple selves emerge. To review:

1. The Oral Bio-Survival System, largely determined by early infantile imprints, deals with seeking Safe Space and avoiding the Dangerous or Alien.

If somebody points a gun at you, whatever self has predominance in this bio-survival system "takes over" the brain at once. Whether you run, or faint, or smash the assailant with a karate chop, you won't remember making the decision. "I just found myself doing it," you will say afterward, because the ancient reptilian circuits of this system move as instant reflexes.

2. The Anal Territorial System, largely determined by imprints at the toddler stage, deals with seizing territory and holding some defined status in the mammalian pack or human family and/or community.

Even a low status, once imprinted, will automatically function thereafter and seem "normal". E.g., the persons with an imprinted Bottom Dog self in this system will feel very uncomfortable, insecure and angry if circumstances force them into a sudden Top Dog position...just as automatically as those with imprinted Top Dog selves will feel uncomfortable, insecure and angry if forced into a Bottom Dog position.

3. The Semantic Time-Binding System, imprinted when language and other symbolisms begin to "make sense" to the growing child, deals with speech, thought (internal speech) and making maps and models of the environment.

Since information increases logarithmically, this system tends to produce new maps and models faster and faster as time passes. These new reality-tunnels unleash new technologies, which alter politics, economics and social psychology in unpredictable ways.

While the first two systems maintain the constants of evolution, the semantic system unleashes fractal "chaos" — the mathematical term for high unpredictability. *A time-binding semantic organism, such as the human, departs from evolutionary norms and functions as a revolutionary agent*...at least potentially.

To prevent the accelerated change and Unknown Results of rapid information flow, most societies dim the time-binding function by setting heavy taboos on speech, writing and other communications. Once these taboos began to break down — after the English and American Bills of Rights became widely copied — information flow increased markedly and the world began quantum-jumping from one reality-tunnel to another with dizzying rapidity. This so alarms conservatives (neophobes) that undoing the Bill of Rights has always played a central role in any conservative program.[1]

4. The Socio-Sexual System, imprinted at puberty, produces a characteristic Sex Role and the "self" capable of playing that Role consistently. "Morals" get conditioned on top of this imprint and produce the gradual "civilizing" process by which loyalty to the family can grow into loyalty to any member of the tribe, to higher loyalties to nation-states etc. and even, in recent times, to an emerging

---

[1] Under the Reagan administration, Colonel Oliver North drew up the FEMA plan, allowing the President to suspend the Bill of Rights at whim. Similar plans appear whenever conservatives gain power anywhere.

sense of loyalty to the human species and to the living Earth itself.

As mentioned earlier, four later systems seem to appear frequently in minorities and may play a larger role in our future evolution. We will now describe these emerging Futurist systems.

5. The Neurosomatic System, containing the brain-neuropeptide-immunological feedbacks discussed in our section on "mind/body" unity.

This system has existed long enough, and techniques of activating it have appeared in so many varieties of yoga, shamanism, hypnosis, "faith healing", etc. that almost everybody outside the A.M.A. and CSICOP knows a little about it and folk-lore contains many proverbs relating to it.

The statistics from Prof. Barefoot of Duke University, showing that optimists outlive pessimists, surprise nobody but Fundamentalist Materialists. Folk-wisdom knows enough about neurosomatic feedbacks — without knowing the scientific details at all — that most people "try to cheer the patient up" with some assurance that positive thinking expresses more than wishful thinking and will have some effect on recovery rates.

An evolutionary/revolutionary turning point, or quantum jump, seems imminent (i.e., will probably occur before the year 2000) because scientific study of immunological/ neuropeptide feedbacks, neurochemistry, Ericksonian and post-Ericksonian hypnosis and Neurolinguistic Programming (NLP) seems likely to produce a "scientific yoga" or, as I elsewhere call it, a **HEAD** Revolution — Hedonic Engineering **A**nd **D**evelopment. The neurosomatic healings and neurosomatic "highs" (yogic or chemical ecstasies) found intuitively or accidentally in the past will then give way to a precise technology of staying High and living Well.

Whole magazines already exist devoted to popularizing the latest scientific findings in neurosomatics. A vast public already knows much about the drugs (legal and otherwise), the vitamins, the nutrients, the brain machines and the computer games that allow access to neurosomatic states. This public of HEAD explorers will grow in the next decade, just as the information explosion in the relevant sciences *will unleash newer and better technology to unchain us from the bondage of imprinting and open the gates to metaprogramming (selective re-imprinting).*

The neurological part of this system seems centered in the right brain hemisphere (which explains why most verbalizations about it, until recently, have sounded like gibberish. Elegant verbalisms only emerge after information has passed through the left hemisphere semantic circuits.)

6. The Metaprogramming System, based on yoga and scientific method, began to emerge in the West after the scientific revolution, among various "Hermetic" societies, c. 1500-1700 e.v. It accelerated in the 1960s when LSD showed the majority of psychologists and neuroscientists that rapid changes in human brain functioning could occur easily, given the right techniques. When the government banned LSD, the research moved into "legal" areas — other drugs (some of which the government then added to the Tabu list), isolation tanks, bio-feedback, etc.

Information flow in this system also seems destined to continue accelerating, just as the 'audience" or "consumers" for this information seem to grow exponentially every decade.

To put it simply, Dr. Timothy Leary sounded like a nut (to most people) when he said, nearly 30 years ago, *"You can change your self as easily as you change the channel on a TV."* Now, even though Dr. Leary still suffers from media slander and misrepresentation, the *avant* one-third of the population understands very well what Leary meant, *viz* —

A. *No "essential self" or static ego exists;*
B. *We can meta-program our nervous systems for a variety of "selves", many of them evolutionarily far in advance of the present Terran average.*

As the technology and inner arts of metaprogramming advance, another evolutionary quantum jump will occur, even more profound than the mastery of the neurosomatic system, which will "only" give us Longevity.

Meta-programming will give us Higher Intelligence.

(The neurological part of this system seems located in the frontal lobes — the newest part of the brain.)

7. The Morphogenetic System contains the "selves" and information banks of all living beings. The first descriptions of this system appear in the language of the "reincarnation" model, as the shamans and yogis who imprinted this system could only talk-and-think about the flood of non-ego information by assuming some transcendental ego that jumped across time from one body to another.

Freud and Jung did a little better. Encountering information from this system in the dreams of their patients, they posited a "racial memory" or "collective unconscious". Neither term qualifies as operational science, but the Freudian and Jungian records at least alerted other psychologists to pay attention to non-ego information systems.

LSD, again, accelerated progress. Finding that vast floods of non-ego information from past ages appeared in LSD sessions, Leary, at Harvard, posited a "neurogenetic circuit". Grof, in Czechoslovakia, posited a "phylogenetic unconscious" and other researchers made up other labels or just recorded the data without trying to name it.

The first scientific model of this system appeared in Dr. Rupert Sheldrake's *A New Science of Life*. Where Leary and Grof, like Jung and Freud, assumed the non-ego information, not known to the brain, must come from the genes, Sheldrake, a biologist, knew that genes cannot carry such information. He therefore posited *a non-local field, like those in quantum theory, which he named the morphogenetic field.* This field communicates between genes but cannot be found "in" the genes — just as Johnny Carson "travels" between TV sets but cannot be found "in" any of the TV sets that receive him.

It will probably take a long, long time — maybe a quarter of a century (i.e., not until around 2015) before we learn the art and science of using the morphogenetic system for fun and profit.

Nonetheless, those who have the most experience of this system all seem to agree with Jung (and Leary): this information system contains not only memories of the past but distinct trajectories of the future.

The morphogenetic system may serve as a kind of evolutionary "radar" *preparing us for future quantum jumps in consciousness by showing us the records of past mutations.*

8. The Non-Local Quantum System (described by modern physics in the last chapter) appears in the reports of a few shamans, yogis and poets in almost every century since the dawn of history. Parapsychologists have made the beginnings of a scientific study of how this non-spatio-temporal system interacts with our other "selves", but largely lacked the operational vocabulary to make their work precise and scientifically crisp. The recent developments in quantum mechanics now open the way to much more rapid progress in understanding "paranormal" and "transcendental" states.

When the "self" operates on the non-local system it becomes a different "self" again, just as always happens whenever we move up from one of these systems to another. The non-local "self" — beyond time and space — and also beyond "mind" and "matter" — has not yet survived translation into left–brain linear verbalism. It transcends all either/ors and, as Buddhists know, we cannot even properly call it a "self".

The Chinese, who seem to have had more experience with this system than anybody else (more than the Hindus, even) define non-local experience in negatives — "not mind", "not self" "not doing", "not existence", even "not non-existence".

The same super-synergy appears in Dr. Bohm's attempts to describe his implicate order in words. However clear his math, his words begin to sound Chinese when he says *the implicate order does not consist of "mind" but that it has "mind-like qualities."*

Obviously, it will take us at least 50 years to get a scientific handle on this level of quantum psychology.

Meanwhile, we at least have learned from the Copenhagenists that whatever model we make of non-local experience, the model will always contain less than the experience itself.

That should save us from the dogmatism, and the gibbering incoherence, of most writers who have tried to discuss the non-local Self.

I would consider it the height of intellectual laziness and mental incompetence to invoke the word "God" to cover the limitations of my imagination and vocabulary. Instead, I will conclude with the wise words of Aleister Crowley. When asked to define the Tao he said,

The result of subtracting the universe

from itself.

New Falcon Publications

Invites You to Visit Our Website:
http://www.newfalcon.com

At the Falcon website you can:

- Browse the online catalog of all of our great titles
- Find out what's available and what's out of stock
- Get special discounts
- Order our titles through our secure online server
- Find products not available anywhere else including:
    - One of a kind and limited availability products
    - Special packages
    - Special pricing
- Get free gifts
- Join our email list for advance notice of New Releases and Special Offers
- Find out about book signings and author events
- Send email to our authors (including the elusive Dr. Christopher Hyatt!)
- Read excerpts of many of our titles
- Find links to our author's websites
- Discover links to other weird and wonderful sites
- And much, much more

Get online today at http://www.newfalcon.com

FROM ROBERT ANTON WILSON

COSMIC TRIGGER I
Final Secret of the Illuminati

The book that made it all happen! Explores Sirius, Synchronicities, and Secret Societies. Wilson has been called "One of the leading thinkers of the Modern Age."

"A 21st Century Renaissance Man. …funny, optimistic and wise…"
                        —*The Denver Post*

ISBN 1-56184-003-3

COSMIC TRIGGER II
Down to Earth

In this, the second book of the *Cosmic Trigger* trilogy, Wilson explores the incredible Illuminati-based synchronicities that have taken place since his ground-breaking masterpiece was first published.

Second Revised Edition!

"Hilarious… multi-dimensional… a laugh a paragraph."   —*The Los Angeles Times*

ISBN 1-56184-011-4

COSMIC TRIGGER III
My Life After Death

Wilson's observations about the premature announcement of his death, plus religious fanatics, secret societies, quantum physics, black magic, pompous scientists, Orson Welles, Madonna and the Vagina of Nuit.

"A SUPER-GENIUS... He has written everything I was afraid to write."
                        —Dr. John Lilly, psychologist

ISBN 1-56184-112-9

FROM ROBERT ANTON WILSON

PROMETHEUS RISING

Readers have been known to get angry, cry, laugh, even change their entire lives. Practical techniques to break free of one's 'reality tunnels'. A very important book, now in its *eighth* printing.

"*Prometheus Rising* is one of that rare category of modern works which intuits the next stage of human evolution... Wilson is one of the leading thinkers of the Modern age."
—Barbara Marx Hubbard

ISBN 1-56184-056-4

QUANTUM PSYCHOLOGY

How Brain Software Programs You & Your World

The book for the 21st Century. Picks up where *Prometheus Rising* left off. Some say it's materialistic, others call it scientific and still others insist it's mystical. It's all of these—and none.

Second Revised Edition!

"Here is a Genius with a Gee!"
—Brian Aldiss, *The Guardian*

"What great physicist hides behind the mask of Wilson?"
—*New Scientist*

ISBN 1-56184-071-8

FROM ROBERT ANTON WILSON

ISHTAR RISING
Why the Goddess Went to Hell and What to Expect Now That She's Returning

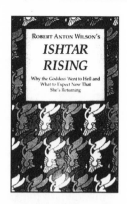

The Return of the Goddess. Wilson provides a new slant on this provocative topic. Exciting, suggestive, and truly passionate. First published by Playboy Press as *The Book of the Breast*. Updated and revised for the '90s. All new illustrations.

ISBN 1-56184-109-9

SEX, DRUGS & MAGICK
A Journey Beyond Limits

Both Sex and Drugs are fascinating and dangerous subjects in these times. First published by Playboy Press, *Sex and Drugs* is *the* definitive work on this important and controversial topic.

"Wilson pokes and prods our misconceptions, prejudices and ignorance."
— Ray Tuckman, *KPFK Radio*

ISBN 1-56184-001-7

THE NEW INQUISITION
Irrational Rationalism & The Citadel of Science

Wilson dares to confront *the* disease of our time which he calls 'Fundamentalist Materialism'. "I am opposing the Fundamentalism, not the Materialism. The book is deliberately shocking because I do not want its ideas to seem any less stark or startling than they are..."

ISBN 1-56184-002-5

FROM ROBERT ANTON WILSON

WILHELM REICH IN HELL

*Foreword by C. S. Hyatt, Ph.D.
and Donald Holmes, M.D.*

Inspired by the U. S. government seizure and burning of the books and papers of the world famous psychiatrist Dr. Wilhelm Reich. "No President, Academy, Court of Law, Congress, or Senate on this earth has the knowledge or power to decide what will be the knowledge of tomorrow."

"Erudite, witty and genuinely scary..."   —*Publishers Weekly*

ISBN 1-56184-108-0

COINCIDANCE

A Head Test

The spelling of the title is *not* a mistake. *Dance* through Religion for the Hell of It, The Physics of Synchronicity, James Joyce and Finnegan's Wake, The Godfather and the Goddess, The Poet as Early Warning Radar and much much more...

"Wilson managed to reverse every mental polarity in me, as if I had been pulled through infinity."
                    —Philip K. Dick, author
                        of *Blade Runner*

ISBN 1-56184-004-1

FROM ROBERT ANTON WILSON

REALITY IS WHAT YOU CAN GET AWAY WITH

What is Humphrey Bogart doing in a movie with Popeye, George Bush and Elvis Presley? That's what the archaeologists of the future are trying to puzzle out. Now you, too, can inspect this outrageous cinematic fabrication set in a time very like our own ...

"A fun romp ... the best screen is inside your head, just waiting to have this rollicking adventure projected on it."
—Mike Gunderloy *Factsheet Five*

ISBN 1-56184-080-7

THE WALLS CAME TUMBLING DOWN

"The title refers not only to the walls of Jericho in the Bible fable but also to the tunnel-walls of the labyrinth of Minos in the Greek myth, which hid Theseus and the Monster from each other for a long while before their final confrontation. Of course, I also had in mind the walls of our individual reality-tunnels..."

"With his humorous rapier, Wilson pokes and prods our misconceptions, prejudices and ignorance. A quantum banquet."
—Ray Tuckman, KPFK Radio

ISBN 1-56184-091-2

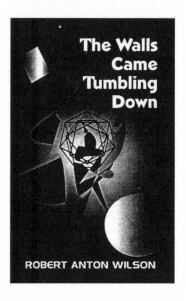